Custom Tasks for SAS® Enterprise Guide® Using Microsoft .NET

Chris Hemedinger

support.sas.com/bookstore

The correct bibliographic citation for this manual is as follows: Hemedinger, Chris. 2012. *Custom Tasks for SAS® Enterprise Guide® Using Microsoft .NET*. Cary, NC: SAS Institute Inc.

Custom Tasks for SAS® Enterprise Guide® Using Microsoft .NET

Copyright © 2012, SAS Institute Inc., Cary, NC, USA

ISBN 978-1-61290-097-1 (electronic book)
ISBN 978-1-60764-678-5

All rights reserved. Produced in the United States of America.

For a hard-copy book: No part of this publication may be reproduced, stored in a retrieval system, or transmitted, in any form or by any means, electronic, mechanical, photocopying, or otherwise, without the prior written permission of the publisher, SAS Institute Inc.

For a web download or e-book: Your use of this publication shall be governed by the terms established by the vendor at the time you acquire this publication.

The scanning, uploading, and distribution of this book via the Internet or any other means without the permission of the publisher is illegal and punishable by law. Please purchase only authorized electronic editions and do not participate in or encourage electronic piracy of copyrighted materials. Your support of others' rights is appreciated.

U.S. Government Restricted Rights Notice: Use, duplication, or disclosure of this software and related documentation by the U.S. government is subject to the Agreement with SAS Institute and the restrictions set forth in FAR 52.227-19, Commercial Computer Software-Restricted Rights (June 1987).

SAS Institute Inc., SAS Campus Drive, Cary, North Carolina 27513-2414

1st printing, December 2012

SAS provides a complete selection of books and electronic products to help customers use SAS® software to its fullest potential. For more information about our e-books, e-learning products, CDs, and hard-copy books, visit **support.sas.com/bookstore** or call 1-800-727-3228.

SAS® and all other SAS Institute Inc. product or service names are registered trademarks or trademarks of SAS Institute Inc. in the USA and other countries. ® indicates USA registration.

Other brand and product names are registered trademarks or trademarks of their respective companies.

Contents

About This Book ix
About The Author xiii
Acknowledgments xv

Chapter 1: Why Custom Tasks 1
Why Isn't Everything Built In for Me? 2
Options for Custom Processes in SAS Enterprise Guide 2
    What Can I Do with Custom Tasks? 4
    Who Uses Custom Tasks Today and What Do They Use Them For? 5
Deploying Custom Tasks 6
    Method 1: Drop-In Deployment 7
    Method 2: Add-In Manager 9
    Accessing Custom Tasks from the Menu 9
    Common Questions about Task Deployment 10
Accessing Ready-to-Use Example Tasks and Source Code 12
    Exercise: Download, Deploy, and Access Custom Tasks from SAS 12
Chapter Summary 12

Chapter 2: Tools of the Trade 15
Introduction to the Microsoft .NET Framework 15
    Which Language Is Right for You? 17
Introduction to Microsoft Visual Studio 18
    Selecting a Version of Microsoft Visual Studio 18
Other Tools to Make You More Productive 21
Learning to be a .NET Programmer 23
    Resources for the New .NET Programmer 23
    Build or Buy: Using Third-Party Components 24
Chapter Summary 25

Chapter 3: Creating Custom Task Projects in Microsoft Visual Studio 27
Overview of the Process 27
    Getting the Project Templates for Microsoft Visual Studio 28
Creating New Custom Tasks Using Project Templates 28
    Using Microsoft Visual Studio 2010 or 2008 29
    Creating a Custom Task Project Using Microsoft Visual Studio 2003 37
Chapter Summary 40

Chapter 4: Meet the Custom Task APIs 41
About Interfaces 41
Meet the Interfaces 43
Understanding the Life Cycles of Your Task 44
Special Interfaces for Special Tasks 49
More Interfaces That Play Nice 52
Chapter Summary 53

Chapter 5: Meet the Task Toolkit 55
Task Toolkit: What's In It? 55
    The SasTask Class 56
    The SasServer Class and SAS.Tasks.Toolkit.Data Namespace 59
    The SAS.Tasks.Toolkit.SasSubmitter Class 60
    The SAS.Tasks.Toolkit.Helpers Namespace 61
    The SAS.Tasks.Toolkit.Controls Namespace 62
    Examples of Using the SAS.Tasks.Toolkit Classes 63
Chapter Summary 67

Chapter 6: Your First Custom Task Using Visual Basic 69
Your First Custom Task Using Visual Basic Express 69
    Creating the Project 70
    Turning the Visual Basic Class into a Custom Task 71
    Build, Deploy, and Test 74
    Adding a User Interface 75
    Saving and Restoring Task Settings 77
Chapter Summary 81

Chapter 7: Your First Custom Task Using C# 83
Your First Custom Task Using Visual C# Express 83
    Creating the Project 84

    Turning the C# Class into a Custom Task 85
    Build, Deploy, and Test 89
    Saving and Restoring Task Settings 92
Chapter Summary 96

Chapter 8: Debugging Techniques: Yes, You Will Need Them 97

Best Practices for Making Your Software Debuggable 98
    Take Advantage of Object-Oriented Design 98
    Consider Unit Testing 99
    Catch and Handle Exceptions 100
    Use Logging to Record Events and Progress 106
Debug with Microsoft Visual Studio 111
    Debugging Basics: Some Definitions 111
    Prepare to Debug a Custom Task 112
    How to Attach a Debugger to a Custom Task 112
    Example: Debugging a Custom Task 113
Chapter Summary 116

Chapter 9: The Top N Report 117

About This Example 118
    Example Source Files and Information 118
Step 1: Exploring the Problem 119
Step 2: Creating the SAS Program 121
Step 3: Creating the Custom Task 123
Examining the Top N Report Solution 123
Chapter Summary 129

Chapter 10: For the Workbench: A SAS Task Property Viewer 131

About This Example 131
    Example Source Files and Information 132
What's in Your Project? 132
    Displaying Properties in a Simple User Interface 134
    Accessing Properties Using the ISASProject APIs 135
    More Possibilities with SAS Enterprise Guide Projects 138
Chapter Summary 139

Chapter 11: Calculating Running Totals 141

About This Example 142

Example Source Files and Information 142
Designing the Task Features 142
    Assumptions: They Are Necessary 143
    Scenario 1: Calculate the Running Total for One Measure across All Rows 143
    Scenario 2: Calculate Running Totals across Groups 144
Designing the User Interface 146
    Assembling the User Interface 147
    Hooking the Controls to Data and Events 151
Saving User Selections 157
    Using LINQ to Create XML 157
    Using LINQ to Read XML 159
Generating a Correct SAS Program 160
    Creating a Readable Program Header 160
    Applying the Task-Specific Filter 161
    Wrap Your Variable Names Appropriately 162
Chapter Summary 163

Chapter 12: The Top N Report 165

About This Example 166
    Example Source Files and Information 166
Dissecting a SAS Data Set 167
Using .NET to Read Data from SAS Data Sets 167
Creating an Elegant Task Flow 169
Adding the SAS Enhanced Editor to a Windows Form 170
Using ISASTaskExecution to Take Matters into Your Own Hands 172
Cancel: Support Is Optional 173
ResultCount: How Many Results? 174
Run, Task, Run! 174
Chapter Summary 176

Chapter 13: Putting the Squeeze on Your SAS Data Sets 177

About This Task 177
    Example Source Files and Information 178
Adapting the Sample 178
    Refactoring a Macro 178
    Compressing the Data Even Further 179
    See How You Did: Adding Reporting 180

Wrapping the Sample in a Task 183
    Designing a User Interface 183
    Modeling the Options in a .NET Class 187
    Putting It All Together: Running and Repeating the Task 189
Chapter Summary 191

Chapter 14: Take Command with System Commands 193

About This Task 193
    Example Source Files and Information 194
Building a Task That Runs Commands 194
    The Structure of This Task 195
    Implementing a Task with ISASTaskExecution 196
Chapter Summary 201

Chapter 15: Running PROCs on Your Facebook Friends 203

Facebook to SAS: The Approach 204
    Example Source Files and Information 204
Gathering Data from Facebook 205
    Example of Transforming JSON to DATA Step Statements 205
Analyzing Data from Facebook 206
    Preparing Data for Reporting 207
    Creating Reports That Provide Insight 208
    Using the Facebook API 209
Running the Example 210
Inside the Task 211
    Connecting to Facebook and Collecting Data 212
    Modeling Data Records with .NET Data Structures 213
    Keeping the User Interface Responsive 216
    Saving the Results in Your SAS Enterprise Guide Task 217
Chapter Summary 218

Chapter 16: Building a SAS Catalog Explorer 219

About This Task 219
    Source Files for This Example 220
About SAS Catalogs 220
    Working with SAS Catalogs 221
The SAS Catalog Explorer Interaction 225
    Creating a Utility Window That Floats 226
    Using SAS Workspace APIs 227

Using Windows Presentation Foundation (WPF) 230
Chapter Summary 231

Chapter 17: Building a SAS Macro Variable Viewer and SAS System Options Viewer 233

About These Tasks 234
    Example Source Files and Information 234
Creating a Productive User Experience 234
    Creating a Toolbox Window 235
    Remembering the Window Position 238
    Adding an About Window with Version Information 241
Designing Code with Objects and Lists 242
    Planning for Object-Oriented Design 243
    Using Data Structures to Represent SAS Objects 244
Performing Other Cool Tricks 247
    Checking the Version of SAS 247
    Running SAS Language Functions 248
    Parsing the SAS Log to Detect Line Types 249
Chapter Summary 251

Index 253

About this Book

Is this Book for You? ix
What You Should Know Before Reading x
How this Book is Organized x
How to Access the Source Code Examples xi
More Resources for Fun and Learning xi

SAS Enterprise Guide ships with a large number of tasks for data manipulation, analysis, and reporting. In addition, it supports a powerful framework for developing business-specific custom tasks. Using this framework with Microsoft .NET technology, you can extend the product to include custom tasks to fit the needs of your site, company, or industry. Plus, the techniques and examples presented in this book also apply to the SAS Add-In for Microsoft Office.

Because SAS Enterprise Guide is accompanies SAS for Windows on desktops, and is deployed with virtually every enterprise-wide SAS business intelligence installation, it becomes a ubiquitous platform – a perfect place to host the custom features that complete your business or analytics offering.

Is this Book for You?

There is no getting around it: this book is intended for programmers. You don't need to have the word "programmer" in your job title, but you do need to be comfortable with writing code, testing code, and debugging.

There are two types of programming described in this book: SAS programming and .NET programming. And if you are a programmer, then you know that programming in one language does not automatically make you an expert in *all* programming languages…especially if one of those languages is SAS. (That is to say, SAS is unlike most other programming languages, both in its syntax and its capabilities.)

You will find this book useful if:

You support a group of SAS users, and you want to make some of their tasks easier and require less programming from them (or from yourself, as a support person)

You work as a consultant, and you want to package some of your proprietary SAS-based methods into an easy-to-use set of tasks.

You serve as IT support for your company, and you want to help the SAS users in your organization to integrate their SAS processes with other business processes that you have in place.

You want to build new tools for yourself, to make your experience with SAS Enterprise Guide even more productive.

What You Should Know Before Reading

This book does not assume that you are an expert in SAS programming. Nor does it assume that you are an expert with .NET programming.

However, this book is also not a tutorial for getting started with SAS or getting started with .NET. There are many other fine publications and online resources for beginning programmers in both of these disciplines.

This book introduces you to the methods that you can use to *combine* SAS programming with .NET knowledge, and to extend your desktop software to help you to use SAS more effectively.

How this Book is Organized

This book is organized into two main sections:

Concepts and Tools

The chapters in the first part of the book will introduce you to the concepts behind custom tasks, the tools that you can use to build them, the programming interfaces (APIs) that SAS provides to support them, and step-by-step instructions for building your first custom task.

The "concepts" material is covered in Chapter 1 through Chapter 8.

Custom Task Examples

The chapters in the second part of the book provide a variety of examples that are ready to use. You can also examine the complete source files (available online) for each example to learn how it works and to tweak the example for your own use.

Each example illustrates different parts of the custom task APIs and different .NET programming techniques.

How to Access the Examples and Source Code

In addition to the gripping prose that is presented within the chapters of this book, you are also entitled to enjoy and learn from a number of actual custom task projects.

Each custom task project is contained within its own source folder, already organized as a project that you can use in Microsoft Visual Studio. Most of the projects feature source code in the C# programming language. A few of the projects also have Visual Basic .NET projects as well.

The project source code, as well as the compiled, ready-to-use binaries, can all be found as links from the companion web site for this book: http://support.sas.com/hemedinger.

Software Used to Develop the Examples

You can use this book with a wide range of software releases from SAS and Microsoft. Most of the examples were developed for SAS Enterprise Guide 4.2 and later, or the SAS Add-In for Microsoft Office 4.2 and later. (At the time of this writing, these include versions 4.2, 4.3, and 5.1.) All of these examples should work with SAS 9.2 and later. A few examples are compatible with SAS 9.1.3.

The Microsoft software includes Microsoft .NET Framework, ranging from version 1.1 through 4.0. Each version of .NET is appropriate for a different release of the SAS software, and each uses a different version of Microsoft Visual Studio. Chapter 2, "Tools of the Trade" contains a complete list of all of the versions and their compatibility restrictions.

In general, all examples should work correctly with the latest versions of SAS and Microsoft software. Some of the examples will work only with the most recent versions, while other examples are specifically designed to be usable with earlier versions of the tools.

More Resources for Fun and Learning

As with most technical books, the topics presented here are somewhat of a moving target. SAS Enterprise Guide continues to evolve, as does the Microsoft .NET Framework. This book provides a solid base for learning the concepts behind creating your own tasks, but new techniques and APIs are sure to emerge even as you read these words.

For the latest examples and techniques from the author, visit The SAS Dummy Blog at http://blogs.sas.com/sasdummy. You will find timely posts about custom tasks as well as posts about many other SAS-related topics.

You can also learn a lot more about SAS by visiting the SAS support web site at http://support.sas.com. From there you will find useful notes and articles, training courses, and books. You can also connect with other members of the SAS community on the focused discussion forums. The following are some of the direct links that you might find useful.

Knowledge Base	http://support.sas.com/resources/
Support	http://support.sas.com/techsup/
Training and Bookstore	http://support.sas.com/learn/
Community	http://support.sas.com/community/

To learn more about programming with Microsoft .NET, visit the Microsoft developer web site at http://msdn.microsoft.com. You'll find a prominently displayed "Learn" section that contains links to tutorials, books, and examples.

About The Author

Chris Hemedinger has been with SAS since 1993. In that time he has been a writer, a software developer, and an R&D manager. He was on the founding team for SAS Enterprise Guide, leading the development team for 10 years. Today, Hemedinger is a principal technical architect in SAS Professional Services (also known as "consulting"). He is also coauthor of *SAS For Dummies* and writes The SAS Dummy blog at http://blogs.sas.com/sasdummy.

Learn more about this author by visiting his author page at http://support.sas.com/hemedinger. There you can download free chapters, access example code and data, read the latest reviews, get updates, and more.

Acknowledgments

It took me 5 years to complete this book.

Technology changes quickly. There are not many technical topics that you can take 5 years to develop into a book and still have a relevant product at the end. Fortunately, this book is primarily about SAS, and SAS has "staying power" in the tech world.

Even with SAS' stable nature, the past 5 years have spanned 3 major versions of SAS, 4 versions of SAS Enterprise Guide, 4 versions of the SAS Add-In for Microsoft Office, and 5 versions of Microsoft .NET. All of these products (with all of their versions) are addressed in some way within this book, which I hope will make the book and its examples relevant and useful for a long time.

Aside from my own glacial-pace writing effort, many other people have contributed to the quality of this book. John West from SAS Press helped me to start the project, and then eventually he handed it off to Stephenie Joyner to cajole me into finishing it. Amy Wolfe edited the text such that when she was done, I couldn't tell which wording she had changed: the definition of the perfect copyedit. She even tried out the programming examples – that's "above and beyond" for a copyeditor.

The book has also benefited from a number of technical reviewers and contributors:

- Members of the SAS Enterprise Guide development and testing team, including David Bailey, Mike Porter, Stephen Slocum, and Hunter Weaver.
- SAS education and training consultants, including Rick Bell, Neil Constable, Kelly Gray, Martha Hays, Mark Jordan, and Andy Ravenna.
- My "early adopters" who have a keen interest in the topic and had to make do with "working drafts" for quite a long time, including Simon Arnold, Angie Hedberg, Jessie Lan, Daniel Ringquist, and Bill Sawyer. (And I know that there were others of you out there!)
- And finally, the many SAS customers who have expressed their eager curiosity about custom tasks and how to build them. Thank you for your questions at user group events, on my blog, and on the SAS Enterprise Guide discussion forum; these have all helped to shape the content of the book.

On a personal note, I want to thank my patient wife, Gail, as well as my tolerant daughters: Maggie, Evelyn, and Gwen. I know that this project seemed to loom on forever, and it was their flexibility that allowed me to finally complete the content and get it ready for publishing.

Chris Hemedinger

Chapter 1: Why Custom Tasks?

Why Isn't Everything Built In for Me? 2
Options for Custom Processes in SAS Enterprise Guide 2
    What Can I Do with Custom Tasks? 4
    Who Uses Custom Tasks Today and What Do They Use Them For? 5
Deploying Custom Tasks 6
    Method 1: Drop-In Deployment 7
    Method 2: Add-In Manager 9
    Accessing Custom Tasks from the Menu 9
    Common Questions about Task Deployment 10
Accessing Ready-to-Use Example Tasks and Source Code 12
    Exercise: Download, Deploy, and Access Custom Tasks from SAS 12
Chapter Summary 12

For more than 35 years, people have been writing SAS programs to solve business problems, conduct research, and report on data. For almost as long, the people who use those SAS programs have strived to invent ways to make those programs reusable, adaptable to changing business processes, and approachable by non-programmers.

A custom task provides one way to leverage your SAS programs and make them usable by a wider audience. Custom tasks are a hook for extending SAS Enterprise Guide and the SAS Add-In for Microsoft Office—two popular desktop applications that bring the power of SAS to a wide range of users. If you work with SAS users who use these applications as their primary interfaces to SAS, then custom tasks can bring your SAS solutions right to them, without asking them to leave the environment that they know and love.

A custom task is the easiest and most natural way for a user to access a proprietary process. However, a custom task can require a significant investment in time and expertise to create. In most cases, the return on investment—realized in consistency, control, and ease of use—far outweighs the cost of developing the task.

Why Isn't Everything Built In for Me?

If every conceivable SAS process was already built into SAS and accessible at the touch of a button, there wouldn't be any room for customers to add their own intellectual property to the system. (And, SAS consultants couldn't earn a living.)

SAS offers hundreds of modules, called *procedures*, which perform specific analyses, produce reports, and manipulate data. SAS Enterprise Guide provides almost 90 tasks that use the most popular procedures to accomplish much of what customers need to do.

In addition, SAS provides rich programming languages that you can use to do almost anything in the world of data manipulation, reporting, and logical flow. These languages include the DATA step and the SAS macro language. It would be impossible to create a set of tasks that represents everything that you can do with these languages. Why, SAS experts are still inventing new uses for these languages every day! The application of SAS languages varies from company to company, industry to industry, and department to department.

Options for Custom Processes in SAS Enterprise Guide

If you need to customize a process that isn't covered by a built-in task in SAS Enterprise Guide, you have several options.

Write a SAS program.
For a SAS programmer, it's the ultimate in flexibility. Anything that you can do in a DATA step, macro language, or SAS procedure is available to you in a SAS program. What's the problem with this approach? The only person who understands how to write the ultimate SAS program is a SAS programmer. If you have an audience that includes non-programmers, you probably need something more controlled.

Write a SAS program and add prompts.
With SAS Enterprise Guide, you can add placeholder values, called *prompts*, to your project. When you reference one of these values in a SAS program or in any task, SAS Enterprise Guide presents a user-friendly prompt dialog box that collects values from the user. When you attach prompts to a SAS program, you can make the program more usable. There is a wide range of prompt values that address a variety of situations, and the user doesn't need to

understand the content of the program to provide values. However, the user does need to know how to *run* the SAS program. The prompt dialog box, while somewhat flexible, does not offer much interactivity. You can define prompts that include a predefined list of values and simple range checking, but you have little control over how the prompts are presented.

Create a stored process.

A stored process is a SAS program that you store in a central repository. You can add prompts to the program and define permissions to control who can run it. And, you can run the program from various application environments, including SAS Enterprise Guide, the SAS Add-In for Microsoft Office, SAS Web Report Studio, and custom web applications. A stored process is a great way to reuse a SAS program and use it consistently across the enterprise.

With authoring tools like SAS Enterprise Guide, it is relatively simple to create a stored process. However, just like writing a SAS program with prompts, your control of the user experience is limited. You have little control over how the prompts are presented to the user, although you can include a predefined list of values and simple range checking. Because stored processes can leverage the SAS metadata environment, you can build prompts that are dynamic. Dynamic prompts pull values from live data and permit cascading prompt selections.

Create a custom task.

Anything that you can do with a SAS program, you can do with a custom task. And, anything that you can do with a Windows application (which means *the best* in terms of an interactive and responsive user experience), you can do with a custom task. If you need access to data or resources that you cannot reach with a SAS program, a custom task can act as the bridge to bring this information into SAS.

Unlike a stored process, custom tasks work in only two SAS applications: SAS Enterprise Guide and the SAS Add-In for Microsoft Office. If you have to provide the feature to a user of a SAS web application or a user in SAS Web Report Studio, a stored process might be how to go.

Table 1.1 summarizes your programming options and how they rank in usability and difficulty. I don't want to mislead you: creating a custom task is more involved and requires more technical skill than some of the other options. But, the payoff can be well worth the investment. And, the tools and techniques presented in this book can make creating a custom task much easier.

Table 1.1: Comparing Programming Options in SAS Applications

Option	User Skills	Designer Skills	Usability
SAS program	Run or modify SAS program	Program in SAS	None; works only in code view
SAS program with prompts	Run SAS program; point and click to answer prompts	Program in SAS; design prompts	Constrained by prompting technology; only static prompts are allowed
SAS stored process	Point and click to run process and answer prompts	Program in SAS; design prompts; know about SAS metadata	Constrained by prompting technology; dynamic prompts are possible
SAS custom task	Point and click to run task	Program in SAS; program in Microsoft .NET design UI	Almost anything is possible; UI is intuitive; dynamic and responsive experience

What Can I Do with Custom Tasks?

Custom tasks straddle the gateway between two very powerful worlds: the world of your SAS session and the world of your Windows desktop.

Your SAS session provides access to the most advanced analytics and reporting capabilities available in business intelligence applications today. Your Windows desktop, using Microsoft .NET as a framework, provides a truly innovative and rich user experience while accessing data from your desktop, network, or the Internet.

Here are some example uses for custom tasks:

- Create a user interface as a front end for a SAS procedure or SAS technique that you use all of the time, but that doesn't have a built-in task in SAS Enterprise Guide. With the more than 250 procedures that SAS offers plus the extensive programming language, it's no surprise that there are gaps between what the user interface offers and what SAS can actually do for you. (For an example, see Chapter 9, "The Top N Report.")
- Connect to a web service to gather data from an external source and import the data into SAS. (For an example, see Chapter 15, "Running PROCs on Your Facebook Friends.")

- Provide simple access to SAS macro programs used within your organization and make them more approachable to non-programmers. (For an example, see Chapter 13, "Putting the Squeeze on Your SAS Data Sets.")
- Provide a sign-in prompt to a proprietary database or server resource without revealing the SAS code required for this action.

There are a few things that customers often want to do with custom tasks, but currently cannot. These include:

- Automate the SAS Enterprise Guide interface to add to and run parts of your project. You can accomplish some of this with the SAS Enterprise Guide automation model, which is a separate mechanism from the custom task APIs (application programming interfaces).
- In SAS Enterprise Guide 4.1, you cannot use custom tasks to examine project contents and extract information such as SAS programs. However, there are newer APIs in SAS Enterprise Guide 4.2 and later that allow this. (For an example, see Chapter 10, "For the Workbench: A SAS Task Property Viewer.")

Who Uses Custom Tasks Today and What Do They Use Them For?

Who uses custom tasks? What do they use them for? The short answer is that lots of people use custom tasks for lots of things. Even SAS uses custom tasks to prototype new features in SAS Enterprise Guide and to deliver those new features between major releases.

Before Custom Tasks: SAS/AF

Custom tasks are not a new concept in SAS. For years, SAS customers have created customized SAS applications using the SAS application framework (the product SAS/AF). SAS/AF provides an environment to host a full-screen application based in SAS with a custom user interface and full access to the power of SAS.

The objectives of SAS/AF are the same as the objectives of custom tasks: provide a simplified, focused user interface to enable users to perform a task that is specific to their industry, company, or department. Even now, there are companies that have a large investment in SAS/AF applications and their users continue to interact with SAS in this way.

SAS/AF applications were just the ticket when SAS users used to "live" in the SAS Display Manager, which acted as the host for SAS/AF. Today, most SAS users access SAS using other products, such as SAS Enterprise Guide. New SAS users might never see the full-screen environment of the SAS windowing environment. In fact, because many enterprises centralize access to the SAS session on powerful servers (where there is no windowing environment) instead of on user desktops, it's not even possible for users to access the SAS windowing environment. Users must get to SAS using a SAS product, such as SAS Enterprise Guide.

Bringing Custom Tasks to the Desktop

In 2001, SAS Enterprise Guide 2.0 introduced the concept of custom tasks. At that time, SAS documented a simple set of APIs that customers used with Microsoft Visual Basic 6.0. Since then, desktop technology has advanced, but the concept of custom tasks remains the same. Today, the APIs support Microsoft .NET, so customers can select their language of preference. C# and Visual Basic .NET are the most popular. These are modern programming languages with many uses outside of SAS. Learning Microsoft .NET skills can be a career-enhancing move, helping you grow your marketable professional skills.

In the Field: Custom Tasks at Work

SAS includes SAS Enterprise Guide with many of its packaged solutions. In this context, SAS Enterprise Guide is usually intended for advanced or power users and helps them with ad hoc reporting on data within the solution's domain. To provide this help, solutions have custom tasks that perform special functions. Examples of SAS solutions that include custom tasks are SAS Activity-Based Management, SAS Warranty Analysis, SAS Enterprise Miner, and SAS Forecast Server.

SAS users can create their own custom tasks to provide a wide range of features to their internal audiences. Examples of real-world custom tasks include a wizard to aid with clinical trial analysis, a dialog box to facilitate user sign-on with SAS/CONNECT, a task that provides easy access to data by assigning the SAS libraries needed by a user or a group, and tasks that create a set of reports and charts for use within Microsoft Excel.

The pluggable nature of custom tasks makes them a perfect mechanism for SAS to enhance SAS Enterprise Guide between releases. On the *Downloads & Hot Fixes* page of the support.sas.com site, you can find custom tasks to enhance your SAS Enterprise Guide installation. These tasks can help you upload and download SAS data sets to and from your SAS server, register SAS tables within metadata libraries, and create graphs using an interactive interface.

Deploying Custom Tasks

A custom task is shipped in an executable .NET file called an *assembly*. (An assembly usually takes the form of a DLL file.) You can have multiple custom tasks in a single .NET assembly, or you can have one custom task per assembly (or DLL file), depending on your preferences. Keep in mind that it is more convenient to package several related custom tasks together into a single .NET assembly because it makes it easier to share code and implementations.

Once the .NET assembly has been built, deploying and registering the custom tasks to target client machines is simple, and two methods are available. Once the tasks are added, they are available from the **Add-In** option in the **Tools** menu and from the **Task List** option in the **View** menu in SAS Enterprise Guide. In the SAS Add-In for Microsoft Office, you can find custom tasks by selecting **Manage SAS Favorites**.

Method 1: Drop-In Deployment

You do not need to perform a separate registration step for SAS Enterprise Guide or the SAS Add-In for Microsoft Office to recognize your custom task. You simply copy the DLL file to a special folder and SAS Enterprise Guide or the SAS Add-In for Microsoft Office recognizes it automatically the next time you start either product.

The special folder locations are different, depending on which version of the product you have installed.

SAS Enterprise Guide 4.1 and the SAS Add-In for Microsoft Office 2.1

In SAS Enterprise Guide 4.1 and the SAS Add-In for Microsoft Office 2.1, the special folder location is:

```
C:\Program Files\SAS\Shared Files\BIClientTasks\Custom
```

You need to create the `Custom` subdirectory if it doesn't exist. The first part of the path (`C:\Program Files\SAS\Shared Files`) might vary, depending on your installation.

SAS Enterprise Guide and the SAS Add-In for Microsoft Office 4.2, 4.3, 5.1, and later

Beginning with SAS Enterprise Guide 4.2 and the SAS Add-In for Microsoft Office 4.2, there are multiple special folders for storing custom tasks. These folders are specific to each release, which means that the folder names are slightly different.

Each folder location supports different deployment scenarios. For example, if you do not have administrative privileges on your PC, you might not be able to copy a DLL file to the `Program Files` location. As a result, one of the folder locations is in the user profile area, which is an area specific to each user account on the PC. Most configurations permit a local user to place files in his or her user profile area.

For SAS Enterprise Guide, the folder locations are:

- `%appdata%\SAS\EnterpriseGuide\<version>\Custom`
- `C:\Program Files\SAS\EnterpriseGuide\<version>\Custom` (SAS 9.2 installation)
- `C:\Program Files\SASHome\x86\SASEnterpriseGuide\<version>\Custom` (SAS 9.3 installation)

For the SAS Add-In for Microsoft Office, the folder locations are:

- `%appdata%\SAS\AddInForMicrosoftOffice\<version>\Custom`
- `C:\Program Files\SAS\AddInForMicrosoftOffice\<version>\Custom` (SAS 9.2 installation)
- `C:\Program Files\SASHome\x86\SASAddInForMicrosoftOffice\<version>\Custom` (SAS 9.3 installation)

If you want to deploy a single copy of a custom task in both applications, you can copy it to this folder location:

`%appdata%\SAS\SharedSettings\<version>\Custom`

Notes: The %appdata% value is a Microsoft Windows environment variable that maps to your personal profile area, which is an area specific to your user account on the PC. You can add custom tasks to your installation without affecting other users who might share your machine.

If your custom task depends on other .NET assemblies that are shipped with SAS Enterprise Guide or the SAS Add-In for Microsoft Office, you do not need to copy those .NET assembly files to the `Custom` directory. However, if your custom task depends on .NET assemblies that are not provided by SAS (for example, they are from third-party vendors or they are .NET assemblies that you developed), you *do* need to copy those .NET assemblies to the `Custom` directory when you copy the custom task.

With drop-in deployment, the SAS client application searches the designated `Custom` folders and any applicable subfolders on your system. If your custom task includes several DLL files, you can group them into a single subfolder to make the deployment run more efficiently.

Example Deployment Scenarios

Suppose you have a custom task that uses forecasting techniques to predict product inventory needs. You have built this task into a DLL file named SupplyTasks.dll. How should you deploy the task for use?

Scenario 1: A Few Users on Local Machines
The task is used by just a few people in your organization. They use SAS Enterprise Guide 4.3 on their desktop machines. SAS Enterprise Guide was installed using the default settings with the SAS 9.3 Software Depot.

Scenario 1: Approach
Each user can copy the SupplyTasks.dll file to his machine for use. The user opens Windows Explorer and navigates to `%appdata%\SAS\EnterpriseGuide\4.3`. The user creates a new folder named `Custom` and copies the DLL file into it. The next time the user opens SAS Enterprise Guide 4.3, the task appears in the **Tools→Add-In** menu.

Scenario 2: Many Users on Many Machines
The task is very popular, and more people need to use it. The self-service method of allowing users to install it themselves might not scale well with a larger audience. In some cases, the task might need to be installed on a machine that is shared by multiple users.

Scenario 2: Approach
On each machine, copy the SupplyTasks.dll file to the `Custom` folder in the SAS Enterprise Guide installation directory. Open a Windows Explorer and navigate to `C:\Program Files\SASHome\x86\SASEnterpriseGuide\4.3`. Create a new folder named `Custom` (if it does not already exist) and copy the DLL file into it. The next time any user on the shared machine opens SAS Enterprise Guide 4.3, the task appears in the **Tools**→**Add-In** menu.

Method 2: Add-In Manager

The Add-In Manager dialog box lets you select a custom task from any location and register it for use in your SAS applications. To get started:

1. Copy the .NET assembly (and any dependent .NET assemblies excluding those that are shipped with SAS Enterprise Guide or the SAS Add-In for Microsoft Office) to a location on the target machine. If you plan to deploy more than one .NET assembly with add-in tasks, you can group them into a single directory.
2. To start the Add-In Manager:
    - In SAS Enterprise Guide 4.1, select **Add-In**→**Add-In Manager**.
    - In SAS Enterprise Guide 4.2, 4.3, or 5.1, select **Tools**→**Add-In**→**Add-In Manager**. In the SAS Add-In for Microsoft Office 2.1, run `C:\Program Files\SAS\Shared Files\BIClientTasks\4\RegAddin.exe`. (The actual location of this program might vary depending on your configuration.)
    - In the SAS Add-In for Microsoft Office 4.2 or 4.3, run `C:\Program Files\SAS\AddInForMicrosoftOffice\4.2\RegAddin.exe` or `C:\Program Files\SAS\AddInForMicrosoftOffice\4.3\RegAddin.exe`. (The actual location of this program might vary depending on your configuration.)

3. Click **Browse** to find the .NET assembly, and then click **Open**. The Add-In Manager dialog box shows the available add-in tasks in the .NET assembly that you selected.
4. Click **OK** to accept the add-in tasks.

Accessing Custom Tasks from the Menu

After your custom task is registered (by using drop-in deployment or the Add-In Manager dialog box), you can launch the task from either the **Add-In** menu or the **Task List** in the **View** menu in SAS Enterprise Guide.

In SAS Enterprise Guide 4.1, the top-level **Add-In** menu includes all of the custom tasks that you currently have registered. In SAS Enterprise Guide 4.2 and later, the **Add-In** menu is located under the **Tools** menu. Figure 1.2 shows the **Add-In** menu in SAS Enterprise Guide 4.2.

Figure 1.1: The Add-In Menu in SAS Enterprise Guide 4.2

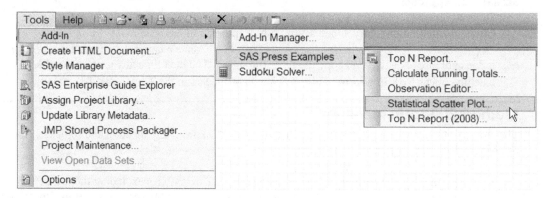

You can also find the custom tasks in the Task List (**View→Task List**). In the Task List, the tasks are organized by category and the custom tasks are intermixed with the built-in tasks. In the **Add-In** menu, the custom tasks are separated into their own Add-In menu structure.

Note: Custom task developers often ask how they can get their custom tasks to appear in the main menus of SAS Enterprise Guide. They want them to be under a top-level menu or in the main categories of Data, Describe, Graph, and so on. Currently, you cannot add a custom task to a built-in menu. Only SAS tasks can appear in these built-in menus. All custom tasks appear under the **Add-In** menu.

Common Questions about Task Deployment

Developers who create custom tasks and the SAS administrators who support SAS Enterprise Guide users often have questions about the best methods to deploy and monitor custom tasks. Here are some common questions with answers:

Can I place the custom task DLL file in a shared network location so that all users can access it without having to copy it to their individual machines?
  The answer is "it depends." In most environments, the Microsoft .NET run time is configured to allow .NET assemblies (DLL files) to load only from trusted locations. Usually, a network location (such as a mapped drive, UNC path, and especially the Internet) is not granted full trust. Without the full trust level, the Microsoft .NET run time refuses to load the .NET assembly.

An administrator can modify the Microsoft .NET security settings to grant levels of trust to a location or to a set of .NET assemblies. The specific instructions are not included in this book, but you find them by searching for "Microsoft .NET code access security" in Microsoft documentation.

Can I control which users are allowed to use my task?
If your SAS environment includes a SAS Metadata Server, then you can control which users have access to your tasks. The SAS metadata environment supports the concept of role-based capabilities.

Each built-in feature in SAS Enterprise Guide is registered as a capability that you can control. You can control who accesses your custom tasks by following these steps:

1. Disable the capability for users to access unregistered custom tasks.
2. For each custom task that you want to add, register the task in the SAS metadata environment. SAS provides the tools and documentation to accomplish this. Each task will appear as a new capability in the SAS metadata.
3. Add the new capability to the role definition of each group or user that you want to allow to access your custom tasks.

The tricky part about relying on metadata roles is that when the user is not connected to a SAS metadata environment, the user's access roles are not checked. If you want your task to be available only when a user has been checked out in metadata, you can code your task as requiring metadata before it is enabled. (For more information, see the SasMetadataRequired attribute in Chapter 5, "Meet the Task Toolkit.")

Can I see which users are actually using my task?
There is no built-in auditing for any of the SAS Enterprise Guide features, including custom tasks. However, there are a few ways to get this information:

- Use the Microsoft .NET logging mechanism (called log4net) to add information about the custom task activity to the SAS Enterprise Guide system log. The logging mechanism is described in Chapter 8, "Debugging Techniques: Yes, You Will Need Them."

- Use Microsoft .NET classes to add information to the Windows event log. The Windows event log is easy to search using built-in administrative tools on Windows. You can even use log4net to add entries to the Windows event log.

- Create your own custom logging mechanism that records when a user uses the custom task. For example, you can use Microsoft .NET classes to add a record to a database or add an entry to a central log that is on the network.

Accessing Ready-to-Use Example Tasks and Source Code

All of the examples that are included in this book are available electronically on the companion web site at http://support.sas.com/publishing/authors/hemedinger.html. The examples are provided as prebuilt .NET assemblies (DLL files), and they are ready to drop in and use with your SAS software. Each example includes one or more Microsoft Visual Studio projects that have all of the source code that you need to build the examples and modify them for your use. Some examples are provided in the C# language, some in Visual Basic .NET, and some in both.

You do not need to understand how these tasks are built before you can use them. You should deploy them in your existing SAS Enterprise Guide installation and explore some of the capabilities that you can add with custom tasks.

Exercise: Download, Deploy, and Access Custom Tasks from SAS

This book includes a collection of custom tasks that illustrate many of the possible features that you can implement. Most of these custom tasks are described in greater detail later in this book. For now, you can deploy a ready-to-use DLL file that contains the tasks and you can see them in action.

To download the collection of custom tasks, follow these steps:

1. Using your web browser, navigate to http://support.sas.com/hemedinger.
2. In the section for this book, click **Ready-to-use custom tasks examples**. A Zip archive file appears. Use your web browser to download the Zip file to your PC.
3. Navigate to the location where the Zip file is located. Extract the contents of the Zip file to a folder on your PC. The Zip file contains a DLL file named SAS.Tasks.Examples.DLL.
4. Copy the DLL file to the appropriate `Custom` folder on your PC based on the version of SAS Enterprise Guide that you are using. For example, if you are using SAS Enterprise Guide 4.3, the location is `%appdata%\SAS\EnterpriseGuide\4.3\Custom`. (For the list of `Custom` folders for each of the different versions of SAS Enterprise Guide, see "Method 1: Drop-In Deployment" earlier in this chapter.)

    Tip: The *%appdata%* location is specific to your local Windows profile area, and the folder structure varies for different versions of Windows. An easy way to navigate to the correct folder is to open a new Windows Explorer and type `%appdata%` in the address bar.
5. Start SAS Enterprise Guide. Select **Tools**→**Add-In**. You should see several new menu items that provide access to the example custom tasks.
6. Select a menu item to launch your first custom task!

Chapter Summary

Custom tasks provide a powerful way to extend SAS business intelligence applications. They aren't the only way to present custom features to your users, but they are often the most flexible way. When you need a custom task, nothing else will do.

Many organizations use custom tasks to add valuable features to SAS Enterprise Guide and the SAS Add-In for Microsoft Office. Even SAS uses custom tasks to enhance its packaged solutions and to provide new features in its products that are in the field.

The mechanism for deploying and using custom tasks is simple. Once it is in place, your users can access custom tasks the same way they access built-in tasks from SAS.

Chapter 2: Tools of the Trade

Introduction to the Microsoft .NET Framework 15
    Which Language Is Right for You? 17
Introduction to Microsoft Visual Studio 18
    Selecting a Version of Microsoft Visual Studio 18
Other Tools to Make You More Productive 21
Learning to be a .NET Programmer 23
    Resources for the New .NET Programmer 23
    Build or Buy: Using Third-Party Components 24
Chapter Summary 25

When you build custom tasks, you are building software. And, as with anything that you decide to build, you are going to need some tools before you can start.

It's an exciting time to be in the software-building business. Thanks in part to the open-source movement, you can dip your toes into the development pool without having to lay out a lot of money first. However, if you decide to get serious about developing custom tasks and you find yourself spending lots of time doing it, you will realize some productivity benefits if you adopt the professional edition, got-to-pay-for-them tools.

Introduction to the Microsoft .NET Framework

The Microsoft .NET Framework is the most recent programming model for creating Windows applications. SAS Enterprise Guide and the SAS Add-In for Microsoft Office are both based in .NET, as are their APIs for custom tasks.

If all you want to do is create some custom tasks, you do not need to understand all of the details about how the .NET Framework works. But, if you *want* to understand all of the details, there are many books that cover the topic in depth, plus resources on the web that can provide fascinating reading material.

Here are the essential points that you should understand about .NET:

You can select your programming language.
Most .NET developers select C# (pronounced see sharp) or Visual Basic .NET. There are more exotic programming languages, such as Python, Ruby, and dozens of others. The examples in this book use C# and Visual Basic .NET. Both of these are object-oriented programming languages. They have similar programming models even though they differ in their syntax structure.

.NET is a managed programming model.
In a *managed* programming environment, you usually don't have to worry about memory management or explicitly allocating and freeing system resources (such as memory). .NET uses a technique called "garbage collection" to detect when objects are no longer needed and can be freed. Java programmers are accustomed to this technique. (This doesn't mean you can be reckless with system resources in your programming. It just means that you don't have low-level memory pointers to deal with.)

In contrast, writing *unmanaged* code means you have to allocate memory and free it when you're done with it. If you have experience with C or C++ programming, and you are entering the world of .NET, the lack of `malloc()` and `free()` routines might make you uncomfortable. You will have to give up some of the control (and responsibility) that you had with C or C++ programming.

Your code is platform independent.
In this case, "platform" refers to any Windows platform that supports .NET, regardless of CPU architecture.

In a .NET development process, regardless of the programming language that you use, all code that you write is compiled down to a Common Intermediate Language (CIL) (sometimes referred to as MSIL (Microsoft Intermediate Language) or IL). It's called an intermediate language because it's not ready to execute on your desktop CPU.

The IL is stored in an assembly, which is in the form of a DLL or EXE file. When it's time to execute your code, the Common Language Runtime (CLR) compiles the IL into CPU instructions that can run on your machine. Because this compilation happens as needed when relevant code paths are traversed, this process is called just-in-time (JIT) compilation.

You have access to a rich set of .NET Framework classes.
Sitting on top of the CLR is a set of thousands of .NET Framework classes that encapsulate user interface controls, data structures, I/O operations, operating system functions, and more.

You can use these classes in your custom tasks to make your programming tasks much easier. As a result, you can concentrate on the content that you are trying to deliver, not on the mechanics of developing a Windows application. All of these classes are accessible from any .NET programming language.

Which Language Is Right for You?

Your choice of programming language does not affect the .NET or SAS features that are available to you. The choice is mostly a matter of personal (or development team) preference. Because all of the code that you write is compiled down to IL anyway, it makes no difference to the CLR how you wrote it in the first place.

If you are familiar with C, C++, Java, or JavaScript, then C# is probably the best fit for you. If you are familiar with Visual Basic or VBScript, then you might be more comfortable with Visual Basic .NET (affectionately known as VB.NET).

This is an unscientific observation—people who are approaching programming problems from an IT support role tend to select Visual Basic .NET. People who create commercial software applications tend to select C#. Visual Basic has a long history in many IT departments, so IT staffers might view Visual Basic .NET as a natural way to proceed. Commercial software applications tend to have roots in C, C++, or Java, so those developers find that C# fits them better.

To give you a flavor of C# and Visual Basic .NET, here is a snippet of code from a custom task presented twice, once in each language. Both of these code examples implement a simple function that opens the Windows **File** dialog box to let the user select a file.

Here is the C# version:

```csharp
public override ShowResult
          Show(System.Windows.Forms.IWin32Window Owner)
{
  // show the Windows File dialog to select a
  // local file to import
  System.Windows.Forms.OpenFileDialog fd =
          new System.Windows.Forms.OpenFileDialog();
  fd.Title = FileDialogTitle;
  fd.Filter = Filter;
  if (fd.ShowDialog(Owner) ==
          System.Windows.Forms.DialogResult.OK)
  {
    fileToImport = fd.FileName;
    System.IO.FileInfo fi =
          new System.IO.FileInfo(fileToImport);
    // change the label in the process flow to
    // reflect the filename
    Label = string.Format("Import {0}", fi.Name);
    return ShowResult.RunNow;
  }
  else
    return ShowResult.Canceled;
}
```

Here is the Visual Basic .NET version:

```
Public Overrides Function _
     Show(ByVal Owner As IWin32Window) As ShowResult
 ' show the Windows File dialog to select a
 ' local file to import
  Dim fd As New OpenFileDialog
  fd.Title = Me.FileDialogTitle
  fd.Filter = Me.Filter
  If (fd.ShowDialog(Owner) = DialogResult.OK) Then
    Me.fileToImport = fd.FileName
    Dim fi As New FileInfo(Me.fileToImport)
    ' change the label in the process flow to
    ' reflect the filename
    MyBase.Label = String.Format("Import {0}", fi.Name)
    Return ShowResult.RunNow
  End If
  Return ShowResult.Canceled
End Function
```

If you study these two examples, you can see the obvious differences in syntax (semicolons and braces in C# versus line endings and `If..End If` structures in Visual Basic .NET). You can also see the similarities (both interact with .NET Framework classes such as `OpenFileDialog` and `String` in the same way).

Introduction to Microsoft Visual Studio

Microsoft Visual Studio is the integrated development environment (IDE) for creating .NET applications. You can create .NET applications without it, but you *wouldn't want to*. The examples in this book are provided as project and solution files for use in Microsoft Visual Studio.

Microsoft Visual Studio provides the environment for designing a user interface, editing code, debugging, and managing projects. It provides many productivity features, such as automatic syntax completion, easy code navigation, and integrated debugging. The time that you save with these features is worth more than the investment that you will make to purchase this tool.

Selecting a Version of Microsoft Visual Studio

There are different versions of Microsoft Visual Studio. The version that you select for your work will depend on what type of custom tasks you need to create, how many of the Professional features you want, and which versions of SAS Enterprise Guide and the SAS Add-In for Microsoft Office you are working with.

Just as SAS produces new versions of its client applications every couple of years, Microsoft does not sit still with its development environments either. At the time of this writing, there are four possible versions of the Microsoft Visual Studio that you can choose from:

- Microsoft Visual Studio 2003, which supports .NET 1.1.
- Microsoft Visual Studio 2005, which supports .NET 2.0.
- Microsoft Visual Studio 2008, which supports .NET 2.0, 3.0, and 3.5.
- Microsoft Visual Studio 2010, which supports .NET 2.0, 3.0, 3.5, and 4.0.
- Microsoft Visual Studio 2012, which supports .NET 2.0, 3.0, 3.5, and 4.5. (At the time of this writing, the 2012 version has not been released, but it is in the final planning stages from Microsoft.)

How should you decide which version to use? The most restrictive variable in this equation is probably the versions of the SAS applications that you want to target. Table 2.1 shows which versions of the .NET Framework are supported in various SAS applications.

Table 2.1: SAS Applications and the .NET Framework Versions They Support

SAS Application and Version	.NET Framework Version
SAS Enterprise Guide 3.0*	.NET 1.1
SAS Enterprise Guide 4.1	.NET 1.1
SAS Add-In for Microsoft Office 2.1	.NET 1.1 or 2.0
SAS Enterprise Guide 4.2	.NET 2.0, 3.0, or 3.5
SAS Add-In for Microsoft Office 4.2	.NET 2.0, 3.0, or 3.5
SAS Enterprise Guide 4.3	.NET 2.0, 3.0, or 3.5
SAS Add-In for Microsoft Office 4.3	.NET 2.0, 3.0, or 3.5
SAS Enterprise Guide 5.1	.NET 2.0, 3.0, 3.5, or 4.0
SAS Add-In for Microsoft Office 5.1	.NET 2.0, 3.0, 3.5, or 4.0

* SAS Enterprise Guide 3.0 is not discussed in this book.

If you are developing custom tasks for SAS Enterprise Guide 4.1, then you need to use Microsoft Visual Studio 2003 (which supports .NET 1.1). For the later versions of SAS applications, you should probably use Microsoft Visual Studio 2010 (because it is the latest version at the time of this writing.)

Note: If you create a custom task that uses a version of the .NET Framework that is *higher* than what the SAS application supports, the custom task might not load in the SAS application. For example, if you use Microsoft Visual Studio 2010 to create a task that uses .NET 4.0, that task is not usable in SAS Enterprise Guide 4.3.

The Express Editions of Microsoft Visual Studio

The folks at Microsoft love it when application developers create software that runs on Microsoft Windows. To encourage that behavior, Microsoft has Express editions of its development tools that you can download and use at no cost. These Express editions do not have all of the features that the Professional editions have, but they do have most of what you need to create custom tasks.

There are Express editions of Microsoft Visual Studio 2005, 2008, and 2010. Each Express edition focuses on a specific programming language. For example, you can download Visual C# Express or Visual Basic .NET Express. (The commercial editions of the Visual Studio development tools incorporate all of the programming languages into a single development environment.)

The Express editions have some limitations, especially in the built-in debugger. Microsoft removed many of the options for debugging and breakpoints in the Express editions. This can make it especially challenging to use the debugger with custom tasks. Debugging is covered in more detail in Chapter 8, "Debugging Techniques: Yes, You Will Need Them."

Recommendation: If you plan to develop custom tasks for production use in your organization, I recommend that you invest in the Professional editions of the Visual Studio development tools. The additional features in the Professional editions will save you time and make it easier to support your work. However, for learning purposes, you can use the Express editions to follow along with this book.

There is no Express edition or free edition of Microsoft Visual Studio 2003, so if you are developing custom tasks for SAS Enterprise Guide 4.1, you need to purchase the Professional edition.

The Microsoft Visual Studio Environment

Regardless of which version or edition you select, Microsoft Visual Studio offers a comprehensive development environment in which you can:

- Manage your custom task project. A project is a collection of files that, when compiled, result in a .NET assembly that is ready for use.

- Create and edit code source files, such as C# code or Visual Basic .NET code. The code editor provides syntax colorcoding and IntelliSense syntax completion features to help you be more productive.

- Design and modify user interface components, such as dialog boxes and controls. You can build your user interface using the palette of built-in controls such as buttons, text fields, and combo boxes. Or, you can design your own controls. You can specify behaviors to associate with user interactions such as what happens when the user clicks a particular button.

- Build your project. When you build your project, the Visual Studio development environment reports any syntax errors or warnings. It helps you navigate to the offending sections of code so that you can resolve the problems.

- Debug your project. You can use the built-in debugger to watch your task execute within SAS Enterprise Guide or the SAS Add-In for Microsoft Office. You can set breakpoints and step through your code to observe how your task works and to diagnose any problems.

Figure 2.1 shows the Microsoft Visual Studio 2010 development environment with an example project loaded.

Figure 2.1: The Microsoft Visual Studio Development Environment

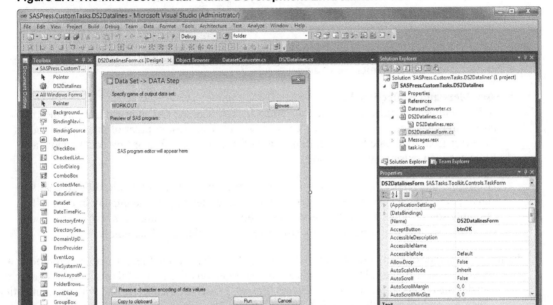

Other Tools to Make You More Productive

Because .NET development is such a big discipline, there are plenty of third-party tools that are designed to make you more productive. Some are packaged as add-ins to Visual Studio and others can be used stand-alone.

These tools include:

Profilers
A profiler is designed to find performance bottlenecks in your code. A profiler instruments your .NET assembly and detects which sections of code take a long time to complete. Profilers identify which sections of code are called repeatedly (making them good candidates for optimization).

A profiler works as your code runs, which makes it different from a static code analyzer.

Static Code Analyzers

A static code analyzer examines your .NET source code and evaluates it using a set of rules. The rules are designed to prevent code errors or exceptions and to enforce good coding practices. The .NET Framework SDK (software development kit) includes a basic set of static code analysis tools called FxCop.

Unlike profilers, static code analyzers do not dynamically evaluate the performance of your code as it runs. A static code analyzer looks only at the source code, performing a static analysis.

The Professional editions of Microsoft Visual Studio provide static code analysis and basic profiling in the development environment.

Disassemblers and Reflection Tools

A disassembler or reflection tool takes a compiled .NET assembly and reconstitutes the IL instructions into source code that you can easily read. The most popular of this type of tool is Reflector from Red Gate Software. Red Gate Software offers a free or low-cost version, and Reflector is an extremely useful tool.

Disassembling a .NET assembly into its source code can seem a bit like cheating. It enables you to peek at the answers that are provided by another developer or company in their .NET assembly source code. Yet, it can serve as a safety net for recovering your own source code when you lose track of it because you still have the compiled .NET assembly.

Note: As an author of many proprietary .NET assemblies, I have mixed feelings about promoting a tool such as Reflector. However, it's a tool that I personally find useful to see what's going on under the covers in the .NET Framework classes. If it is used honestly and with good intentions, then it's an indispensable tool that will help you with your own programming.

Obfuscators

An obfuscator is quite the opposite of a tool like Reflector. An obfuscator takes your .NET assembly and jumbles it so that any downstream user who attempts to disassemble it will have a difficult time learning anything useful. Obfuscation doesn't change the function of your code, it just makes it essentially unreadable. Those who sell third-party components for reuse often obfuscate their .NET assemblies to protect their intellectual property.

These tools can increase your productivity. If you work on a team of several developers, these tools can help your team retain high standards in its coding practices.

If you develop custom tasks for resale, these tools can help you achieve your commercial goals, protect your intellectual property, and improve your code quality. All of these make your tasks less expensive to support!

Learning to be a .NET Programmer

If you are building custom tasks to use in SAS products, then the chances are good that you're already a SAS programmer. But, as part of this exercise, you're about to become something else as well—a .NET programmer.

Most SAS users are aware of the many resources available to help new SAS programmers learn SAS tools and the SAS programming language. There are discussion forums such as communities.sas.com, mailing lists such as SAS-L, and user-supported community sites such as sasCommunity.org.

You will be happy to learn that the .NET community is also very active. On the one hand, its members are not as famously friendly as members of the SAS community (although that's a very high bar). On the other hand, there is plenty of help for new and experienced .NET programmers.

Resources for the New .NET Programmer

Learning styles vary, but .NET programming is such a common discipline that you won't have any trouble finding the resource that best fits your learning style.

Books

There is no shortage of books on the topic of .NET programming. There are many books about .NET development in general, as well as books that focus on specialty topics such as C# or Visual Basic .NET, Windows Presentation Foundation (WPF), architecture, and more. Depending on how sophisticated you decide to make your custom tasks, any of these topics might interest you. The one major area of .NET development that probably won't interest you is ASP.NET, which focuses on web-based development. When you create custom tasks for SAS Enterprise Guide, most of the development is desktop-based and uses Windows Forms or WPF.

In addition to the well-known online bookstores, you can usually find excellent books at your local bookstore if there is a computer technology section. You might even be able to find books at your local public library.

Web Sites

If you like to do your learning on the Internet, here are a couple of websites to get you started:

- **msdn.microsoft.com/net** is the launch point for all things .NET. From this website, you can find links to books, training, virtual labs, tools, and more.
- **www.codeproject.com** is an ad-supported community that contains dozens of discussion forums about the different aspects of Windows programming. This website has a special focus on Microsoft .NET. There are thousands of articles with examples, techniques, and source code that you can reuse in your own applications.

Training
> If you learn best in a classroom in a hands-on environment, there are many trainers certified in Microsoft who would be happy to come to your site and instruct. Or, you can find a class at a training center. The msdn.microsoft.com website is a good place to search for training.
>
> When shopping for training, be sure to select courses that focus on the aspects of .NET programming that you need for custom tasks (for example, C# or Visual Basic .NET, or Windows Forms or WPF). Be sure to determine whether the focus is on creating code libraries or building full applications.

Build or Buy: Using Third-Party Components

Building user interfaces with Microsoft .NET can be fun. There isn't much that you can't do with the technology if you invest enough time to learn and develop with it. However, if you're reading this book, then your main job is probably not .NET developer. You probably also support users of SAS technologies.

As the popularity of Microsoft .NET has grown, so has the industry of supplying components to use with Microsoft .NET applications. If you need a sophisticated user interface control that isn't provided out of the box with the .NET Framework, chances are good that there are one or more third-party vendors who can provide you with just the thing that you need. For example, there are companies that offer specialized data grid components, charting components, and components that integrate with popular document formats such as Microsoft Excel or Adobe PDF.

In addition to third-party components, there are hundreds of open-source projects that provide popular features.

Here are a few web sites that offer code examples and components that you can use. Many of these are open source. You should review the license agreements thoroughly before incorporating the code or libraries into any of your applications.

CodeProject (www.codeproject.com)
> The CodeProject website is devoted almost entirely to Microsoft technologies. Most of the recent content is dedicated to .NET code examples.

CodePlex (www.codeplex.com)
> CodePlex is an open-source project repository sponsored by Microsoft.

SourceForge (www.sourceforge.net)
> SourceForge is a huge repository of open-source projects. A subset of these projects are related to .NET.

GitHub (www.github.com)
> GitHub is a collaborative environment for code projects. It contains thousands of project repositories in the Git version control system that cover all coding technologies including Microsoft .NET. Newer releases of Visual Studio support add-ins that interact with GitHub projects.

Chapter Summary

You can start developing in Microsoft .NET without having to purchase any tools. The Express editions of Microsoft Visual Studio and the many free tools and open-source components enable you to dip your toes into the development pool without any investment except for time.

Software development *is* a discipline, and Microsoft .NET is a dominant subdomain for developing desktop software applications. If software development is a primary part of your profession, it pays to invest in professional tools that ensure professional results.

Chapter 3: Creating Custom Task Projects in Microsoft Visual Studio

Overview of the Process 27
    Getting the Project Templates for Microsoft Visual Studio 28
Creating New Custom Tasks Using Project Templates....................................28
    Using Microsoft Visual Studio 2010 or 2008 29
    Creating a Custom Task Project Using Microsoft Visual Studio 2003 37
Chapter Summary 40

This book assumes that you are using Microsoft Visual Studio as the primary development tool for creating custom tasks. In this chapter, you will learn how to create the basic project structure that enables you to build tasks within Microsoft Visual Studio.

The exact steps that you will follow depend on these factors:

- Which version of Microsoft Visual Studio are you using? The answer to this question depends on which version of SAS Enterprise Guide you are using. See the section "Introduction to Microsoft Visual Studio" in Chapter 2 for more information.
- What is your preferred programming language? The steps are slightly different for Visual Basic .NET versus C#.

The steps in this chapter use project templates (which are extensions for Microsoft Visual Studio) that you can download from support.sas.com. However, you might want to learn how to create custom tasks from the ground up without the help of a template. The step-by-step instructions for doing so are in Chapter 6, "Your First Custom Task Using Visual Basic," and in Chapter 7, "Your First Custom Task Using C#."

Overview of the Process

Regardless of the Microsoft Visual Studio version or preferred programming language, the general steps to create a custom task are the same.

1. Create a project in Microsoft Visual Studio. (The project is a specialized type of class library and uses the project templates available at support.sas.com.)
2. Create the .NET classes to implement the individual task interfaces and the task graphical user interface. Use the templates to help get you started.
3. Create the user interface logic, SAS program, and serialization code. (The serialization code saves and restores task settings from run to run).

Details about each of these steps are provided in this chapter. The steps assume that you are using the project templates and have them installed on your machine.

Getting the Project Templates for Microsoft Visual Studio

SAS provides the project templates. A project template helps you create the shell of your custom task quickly and without having to deal with the mechanics of creating a new project, setting the necessary properties, and adding all of the files that you'll need. With a template, you can bootstrap the beginning of your project. That way, you'll be free to jump right into creating your content.

You can download the project templates from support.sas.com. You can directly navigate to the template files by clicking this direct link:

```
http://go.sas.com/customtasksapi
```

From this page, you'll find different sets of templates depending on these factors:

- Which version of SAS Enterprise Guide are you using? The templates for a SAS Enterprise Guide 4.1 project (Microsoft Visual Studio 2003) are different from the templates for a SAS Enterprise Guide 4.2, 4.3, or 5.1 project (Microsoft Visual Studio 2008 or 2010).
- What is your preferred programming language? There are different templates that support C# versus Visual Basic .NET.

All of the templates are easy to install. You simply copy the template packages into the folder on your PC where Microsoft Visual Studio is installed. Each template package from SAS includes the documentation that describes exactly how to install them.

Creating New Custom Tasks Using Project Templates

Regardless of the SAS Enterprise Guide version or preferred programming language, using a template is simple. Select the menus in Microsoft Visual Studio to create a new project, select the template for the type of custom task that you want to create, click a few buttons, and you're done.

The result is a shell of a project that contains a valid custom task that is ready to build and run. Of course, the task won't *do* much of anything, yet. But, at least the templates get your project in a solid and valid state—a great beginning for your work.

Note: The terminology for custom tasks is a little bit different in SAS Enterprise Guide 4.1. In SAS Enterprise Guide 4.1 and earlier, what is now called a custom task is more commonly referred to as an add-in because a custom task is a way to add in a custom feature to the product.

This book and most SAS documentation now refers to these product extensions as custom tasks. Custom tasks and add-ins are fundamentally the same thing. You will notice that many of the namespaces in the custom task libraries still reflect the older add-in nomenclature.

Using Microsoft Visual Studio 2010 or 2008

Microsoft Visual Studio templates provide a quick start to creating a development project that has a specific purpose or focus. The templates available at support.sas.com help you create a custom task project. They automate the process of creating a simple project with a single custom task that has a simple user interface.

The templates are designed to work with:

- Visual Studio 2010 Professional (or Enterprise or Team System)
- Visual Basic 2010 Express
- Visual C# 2010 Express
- Visual Studio 2008 Professional Edition (or Enterprise or Team System)
- Visual Basic 2008 Express Edition
- Visual C# 2008 Express Edition

Installing the Templates on Your PC

The VS2008Templates.zip file can be downloaded from:

```
http://support.sas.com/documentation/onlinedoc/guide/customtasks/
index.htm
```

The template definitions are in these two files:

- CS2008SasTemplate.zip (for C# projects)
- VB2008SasTemplate.zip (for Visual Basic .NET projects)

Note: These templates are to be used with the .NET assemblies installed with SAS Enterprise Guide 4.2 and the SAS Add-In for Microsoft Office 4.2. Most custom tasks that you build with these .NET assemblies also run with SAS Enterprise Guide 4.3 and the SAS Add-In for Microsoft Office 4.3. No modifications or rebuilds are necessary.

To install the templates to use in Visual Studio:

1. Navigate to the following folder on the PC on which Visual Studio is installed:
   2008 Edition:
   `%userprofile%\My Documents\Visual Studio 2008\Templates\ProjectTemplates`
   2010 Edition:
   `%userprofile%\My Documents\Visual Studio 2010\Templates\ProjectTemplates`
   Note: %userprofile% is a Windows environment variable that resolves to your personal profile area.
2. Create a new folder named `SAS Custom Tasks` under `ProjectTemplates`.
3. Copy the two Zip files (CS2008SasTemplate.zip and VB2008SasTemplate.zip) to the `SAS Custom Tasks` folder. **Do not** extract the files. The templates should remain as Zip archive files for use by Visual Studio.

Creating a Custom Task Project Using Visual Basic 2010 Express

After the templates are installed, here are the steps to create a new custom task project that uses Visual Basic .NET:

1. Open Visual Basic 2010 Express (or Visual Basic 2008 Express).
2. Select **File→New Project**. The New Project dialog box appears as shown in Figure 3.1.

Figure 3.1: The New Project Dialog Box in Visual Basic

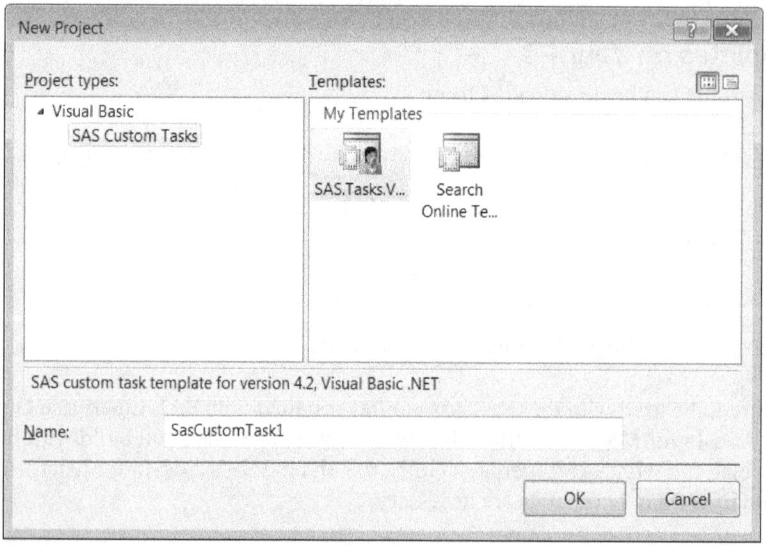

3. Select the SAS.Tasks.VBTemplate42 template.
4. Enter a name for your project.

Chapter 3. Creating Custom Task Projects in Microsoft Visual Studio 31

> **Note:** A one-level name that contains no dot characters works best. For example, use SasStatTask instead of Sas.Stat.Task. Dot characters in the task name can cause namespace problems when the task is built.

5. Click **OK**.

Visual Basic creates a new project with the custom task classes, as shown in Figure 3.2.

Figure 3.2: SasStatistics Visual Basic Project

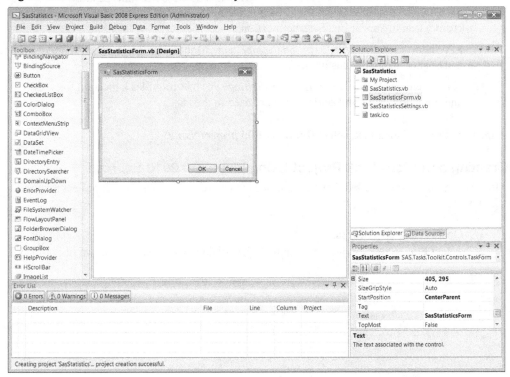

6. Select **File→Save All** to save the new project. The Save Project dialog box enables you to specify a directory location for the new project and its files. When you compile and build the project, the output files appear in the directory location that you specify in this step.

The new project contains:

- A Visual Basic project file with references to the SAS Enterprise Guide DLL files that are needed to build the task. The references will work if SAS Enterprise Guide 4.2 is installed in its default location (C:\Program Files\SAS\EnterpriseGuide\4.2). If SAS Enterprise Guide is installed in a different location (or if it is a different version such as 4.3 or 5.1 or if you are using the SAS Add-In for Microsoft Office), you might need to add a reference path to include the location path for SAS Enterprise Guide or the SAS Add-In for Microsoft Office.

To add or modify a reference path, open the project's Properties dialog box and select **Reference Paths**. Ensure that the correct location is listed and that it is at the top of any other paths that are defined. (As a result, Visual Studio searches that path first.)

- A class file (*projectName*.vb) that implements the SAS custom task APIs. During implementation, the task description information (such as name and category) and the default implementations for saving the task state, launching the task user interface, and generating a SAS program when the task is run are provided. The class file inherits information from the SAS.Tasks.Toolkit.SasTask file, which implements most of the mechanics needed for a custom task to appear in a SAS application.
- A Windows Form class file (*projectNameForm*.vb) that provides a simple Windows Form as a user interface to the custom task.
- A task settings class file (*projectNameSettings*.vb) that provides a location for you to record and track properties and settings that are used in the task.

In the New Project dialog box, select **Build**→**Build** *projectName*.

Creating a Custom Task Project Using Visual C# 2010 Express

After the templates are installed, here are the steps to create a new custom task project that uses Visual C#:

1. Open Visual C# Express.
2. Select **File**→**New Project**. The New Project dialog box appears as shown in Figure 3.3.

Figure 3.3: The New Project Dialog Box in Visual C#

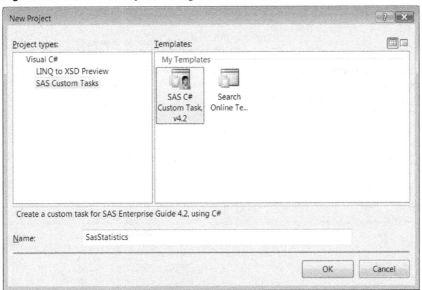

3. Select the SAS C# Custom Task v4.2 template.
4. Enter a name for your project.

   Note: Unlike Visual Basic, the Visual C# environment works well with multilevel namespaces. If you want to create a project with a root namespace such as SAS.Statistics, you can use that as the project name, including the dot character.

5. Click **OK**.

   Visual C# creates a new project with the custom task classes, as shown in Figure 3.4.

Figure 3.4: SasStatistics Visual C# Project

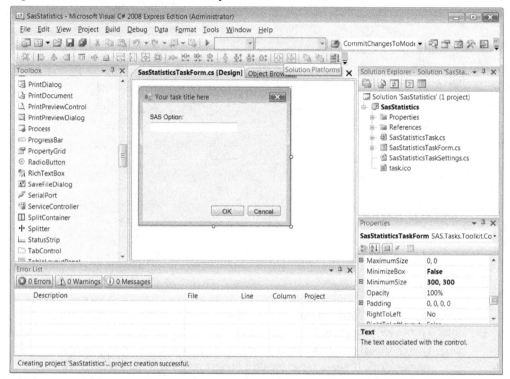

6. Select **File→Save All** to save the new project. The Save Project dialog box enables you to specify a directory location for the new project and its files. When you compile and build the project, the output files appear in the directory location that you specify in this step.

The new project contains:

- A Visual C# project file with references to the SAS Enterprise Guide DLL files that are needed to build the task. The references will work if SAS Enterprise Guide 4.2 is installed in its default location (C:\Program Files\SAS\EnterpriseGuide\4.2). If SAS Enterprise Guide is installed in a different location (or if it is a different version such as 4.3 or 5.1 or you are using SAS Add-In for Microsoft Office), you might need to add a reference path to include the location path for SAS Enterprise Guide or the SAS Add-In for Microsoft Office.

  To add or modify a reference path, open the project's Properties dialog box and select **Reference Paths**. Ensure that the correct location is listed and that it is at the top of any other paths that are defined. (As a result, Visual Studio searches that path first.)

- A class file (*projectNameTask*.cs) that implements the SAS custom task APIs. During implementation, the task description information (such as name and category) and the default implementations for saving the task state, launching the task user interface, and generating a SAS program when the task is run are provided. The class file inherits information from the SAS.Tasks.Toolkit.SasTask file, which implements most of the mechanics needed for a custom task to appear in a SAS application.

- A Windows Form class file (*projectNameTaskForm*.cs) that provides a simple Windows Form as a user interface to the custom task.

- A task settings class file (*projectNameTaskSettings*.cs) that provides a location for you to record and track properties and settings that are used in the task.

In the New Project window, select **Build**→**Build** *projectName*.

Creating a Custom Task Project Using Visual Studio Professional

Templates can be used with Visual Studio Professional (or Enterprise or Team System). The higher-end editions of Visual Studio support many more project types. You can use your preferred programming language (C# or Visual Basic) in this environment.

After the templates are installed, here are the steps to create a new custom task project:

1. Open Visual Studio.
2. Select **File**→**New**→**Project**. The New Project dialog box appears as shown in Figure 3.5.

Figure 3.5: The New Project Dialog Box in Visual Studio

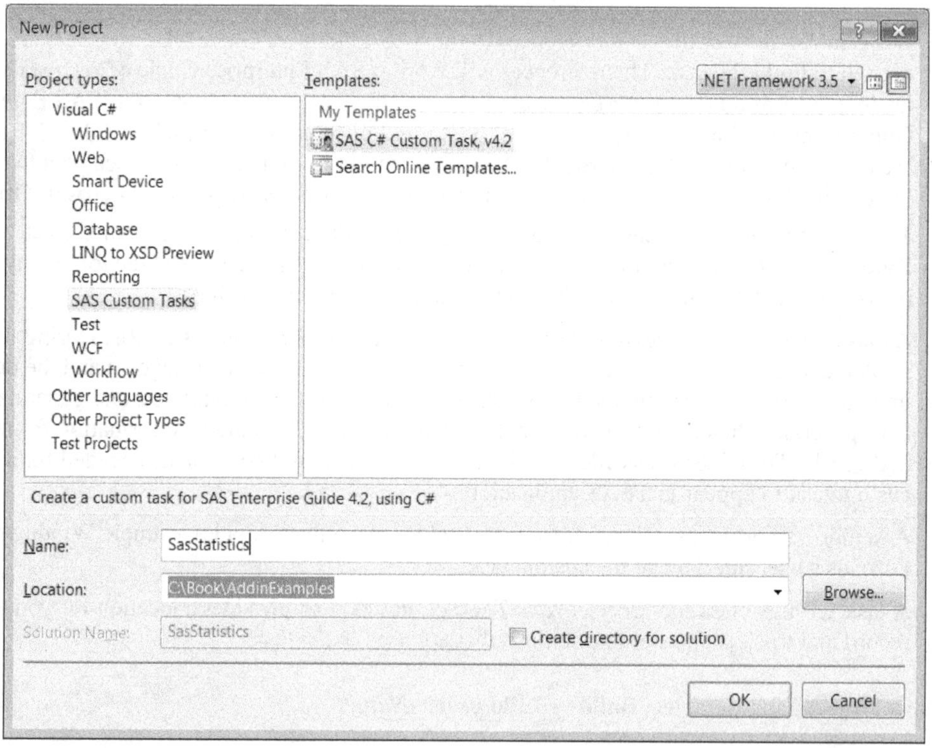

3. Select SAS Custom Tasks in the Project types list. SAS Custom Tasks appears under both Visual C# and Visual Basic (which is under **Other Languages**).
4. Within the SAS Custom Tasks category, select "SAS CustomTask" template. (The template name is slightly different for the C# version and the Visual Basic version.)
5. Enter a name for your project and a directory location in which to store the project's files and to build output.
6. Click **OK**.

Visual Studio creates a new project with the custom task classes.

The new project contains:

- A Visual Studio project file with references to the SAS Enterprise Guide DLL files that are needed to build the task. The references will work if SAS Enterprise Guide 4.2 is installed in its default location (`C:\Program Files\SAS\EnterpriseGuide\4.2`). If SAS Enterprise Guide is installed in a different location (or if it is a different version such as 4.3 or 5.1 or if you are using the SAS Add-In for Microsoft Office), you might need to add a reference path to include the location path for SAS Enterprise Guide or the SAS Add-In for Microsoft Office.

To add or modify a reference path, open the project's Properties dialog box and select **Reference Paths**. Ensure that the correct location is listed and that it is at the top of any other paths that are defined. (As a result, Visual Studio searches that path first.)

- A class file that implements the SAS custom task APIs. During implementation, the task description information (such as name and category) and the default implementations for saving the task state, launching the task user interface, and generating a SAS program when the task is run are provided. The class file inherits information from the SAS.Tasks.Toolkit.SasTask file, which implements most of the mechanics needed for a custom task to appear in a SAS application.
- A Windows Form class file that provides a simple Windows Form as a user interface to the custom task.
- A task settings class file that provides a location for you to record and track properties and settings that are used in the task.

In the New Project window, select **Build→Build** *projectName*.

Creating a Custom Task Project Using Microsoft Visual Studio 2003

If you want to build a custom task for SAS Enterprise Guide 4.1 or the SAS Add-In for Microsoft Office 2.1, then you must use Microsoft Visual Studio 2003. Remember, these versions support only .NET 1.1, which must be built using Microsoft Visual Studio 2003.

After the templates are installed, here are the steps to create a new custom task project that uses Visual Basic .NET or C#:

1. Open Microsoft Visual Studio 2003.
2. Select **File→New→Project**. The New Project dialog box appears as shown in Figure 3.6.
3. Select either **Visual Basic Projects** or **Visual C# Projects**.

Figure 3.6: The New Project Dialog Box in Microsoft Visual Studio 2003

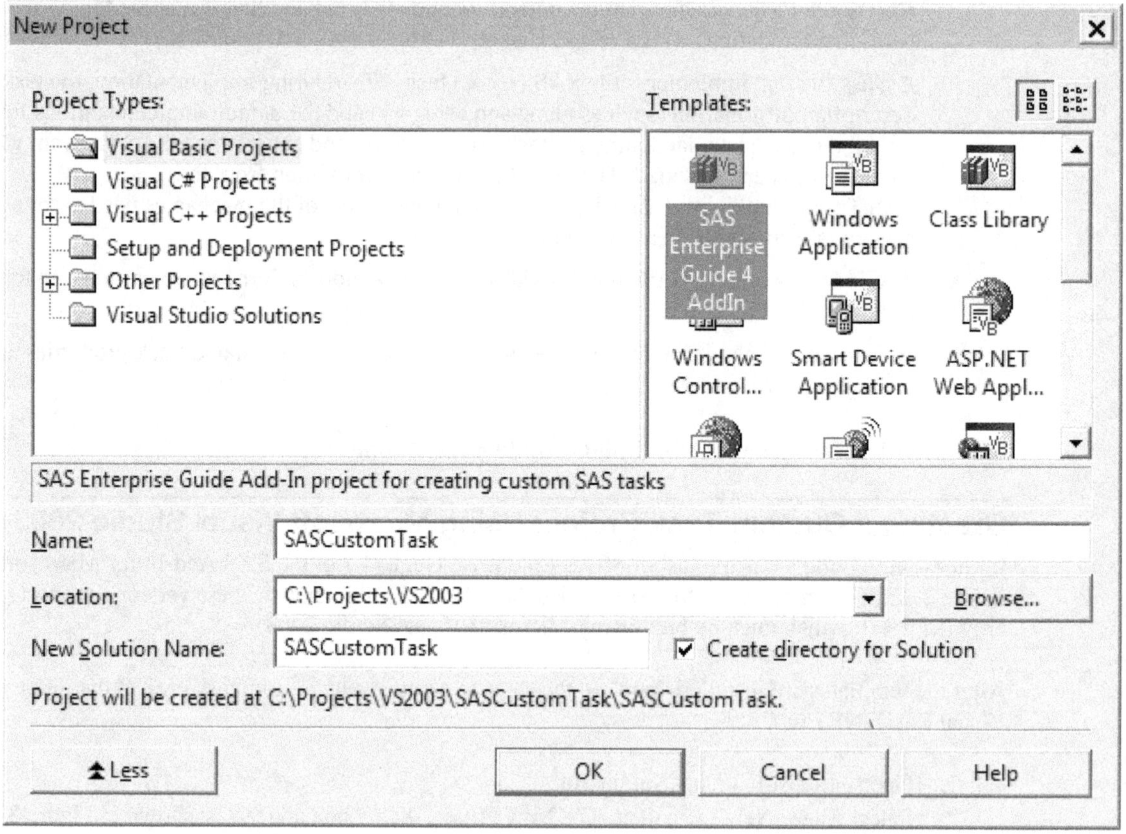

4. Select **SAS Enterprise Guide 4 AddIn**.
5. Enter a name your project and a directory location for the new project and its files.
6. Click **OK**.

   Visual Studio creates a new project with the custom task classes, as shown in Figure 3.7. This project contains all of the references and files needed to build the project.

Figure 3.7: SASCustomTask Visual Studio Project

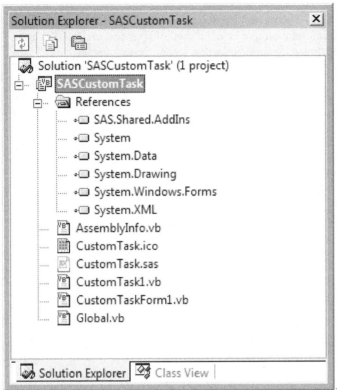

7. From the main menu, select **Build→Build Solution**. Build messages are sent to the output window. (To see the build messages, select **View→Other Windows→Output**.) A successful build will have output that looks something like this:

```
------ Build started: Project: SASCustomTask,
Configuration: Debug .NET ------
Preparing resources...
Updating references...
Performing main compilation...
Building satellite assemblies...
---------------------- Done ----------------------
    Build: 1 succeeded, 0 failed, 0 skipped
```

After the project is built, you can deploy the custom task and then run it. The custom task is saved as a DLL file in a folder named **bin** or **bin/Debug** in your project directory. For more information, see "Deploying Custom Tasks" in Chapter 1, "Why Custom Tasks?"

Chapter Summary

Regardless of the version that you use, Microsoft Visual Studio is a powerful development tool. This chapter provides just enough information to create the shell of your custom task. With this shell in place, you can concentrate on the business logic that makes the task useful.

Chapter 4: Meet the Custom Task APIs

About Interfaces	41
Meet the Interfaces	43
Understanding the Life Cycles of Your Task	44
Special Interfaces for Special Tasks	49
More Interfaces That Play Nice	52
Chapter Summary	53

Building custom tasks involves using APIs and lots of them. APIs are the documented and supported surfaces that enable you to build a custom task that fits into an existing SAS application. You use APIs to make calls to SAS applications, to use .NET components, and to access resources such as data and compute services.

In turn, SAS applications that run your custom tasks use APIs to access your routines and business logic.

As you might guess, APIs play a very important role in building application parts that behave together in a cohesive way. This chapter describes the main APIs associated with custom tasks and their use in SAS applications.

You don't have to know all about custom task APIs to build custom tasks. The SAS Task Toolkit library (available beginning with SAS Enterprise Guide 4.2 and the SAS Add-In for Microsoft Office 4.2) implements most of the nitty-gritty for you, which leaves you to concentrate on the features of your task. Read this chapter now if you want the details. Otherwise, skip to Chapter 5, "Meet the Task Toolkit," and learn about how the Task Toolkit makes your life easier.

About Interfaces

To work together effectively, a custom task and the application that hosts it need a way to communicate. The primary medium for communication is a series of interfaces.

The term *interface* is an overloaded term in the world of computers. In this case, *interface* does not mean user interface. Instead, *interface* means a collection of APIs. These APIs serve as a contract between the application and the custom task. They are an abstract representation of the roles and responsibilities of the application and custom task.

For example, for SAS Enterprise Guide to recognize a custom task as a valid component that can plug into it, the task must implement predictable methods and properties that SAS Enterprise Guide can use. This allows SAS Enterprise Guide to ask questions of the task, such as:

- What's your name?
- Do you require data as input?
- Where can I find your icon to display in my menu?

It allows SAS Enterprise Guide to control the life cycle of the task, with directives such as:

- Initialize yourself.
- Show your user interface.
- Save your state within the project.
- Generate a SAS program to run.

For its part of the contract, SAS Enterprise Guide implements consumer interfaces that provide information and services so that the task can do its job. For example, the task can rely on SAS Enterprise Guide to:

- Provide information about input data sources.
- Provide information about SAS Enterprise Guide options so that it can behave consistently.
- Submit a SAS program on behalf of the task and return the SAS log.
- Provide information about what other items are in the active project.

Because these consumer interfaces are abstract definitions, neither the task nor the application (such as SAS Enterprise Guide) should make any assumptions about *how* the properties or methods are implemented. Interfaces separate the contract from the implementation. For example, the application might ask your task to provide a SAS program to run, but it does not care whether your task generates the program dynamically, returns a pre-built SAS program, or reads from a file on disk and provides the program text.

Both SAS Enterprise Guide and the SAS Add-In for Microsoft Office implement the required consumer interfaces to host custom tasks. Even though the applications operate in different environments and provide different implementations, the fact that they both implement the APIs that serve as the contract means that custom tasks can depend on consistent behavior from both of them.

Meet the Interfaces

To be considered a valid custom task, a task must implement these three interfaces at a minimum:

ISASTaskAddIn
> Provides the basic mechanism for the host application to connect to the task and interact with it as part of the application.

ISASTaskDescription
> Provides information about the task that helps the application surface the task in menus and task lists. In addition, it helps the application decide under what conditions the task can operate (such as checking whether the task requires input data).

ISASTask
> Provides information about a particular instance of the task so that the application can show the task user interface, save and restore the task state, retrieve a SAS program to run, and gather information about expected results and output data.

These three interfaces represent a total of 36 properties and methods that you must implement for even the most basic custom task. Most of the implementation details are simple, but if you create many custom tasks, it can become tedious.

There is good news. Beginning with SAS Enterprise Guide 4.2 and the SAS Add-In for Microsoft Office 4.2, SAS has provided a Task Toolkit library that handles most of the implementation for you. The Task Toolkit is described in Chapter 5, "Meet the Task Toolkit." The Task Toolkit is used in most of the examples in this book.

The host application (such as SAS Enterprise Guide) implements these interfaces:

ISASTaskConsumer
> Provides access to information and features of the host application. It provides access to options in the application that are specific to the particular instance of the task.
>
> The host application provides access to resources and services with some additional interfaces, described next.

ISASTaskData, ISASTaskDataAccessor, and ISASTaskDataColumn
> Provide access to data services so that your task can get information about the active data source.

ISASTaskSubmit
> Enables you to submit a SAS program for processing and retrieve the results (such as the SAS program log) while your task is active.

Versioned Interfaces

Applications don't stand still. As SAS develops and releases new versions of its applications, new capabilities and APIs are introduced. As a result, the behavior of the application might change. Because the APIs serve as a contract between the application and the custom tasks, it's very important that the application preserves and respects the existing APIs as much as possible for the existing tasks, even as the application introduces new capabilities.

As of today, custom tasks that were developed for SAS Enterprise Guide 4.1 should continue to work in SAS Enterprise Guide 4.2, 4.3, and 5.1. Likewise, custom tasks that were developed for the SAS Add-In for Microsoft Office 2.1 should continue to work in the SAS Add-In for Microsoft Office 4.2, 4.3, and later.

To offer new capabilities in new application versions without disrupting the custom tasks built for a previous version, SAS provides the new capabilities in a versioned interface. A versioned interface is usually a superset of the original interface plus new properties and methods that provide enhancements. However, using a versioned interface might still result in a different behavior.

For example, new versions of SAS Enterprise Guide provide more features than those originally offered in the ISASTaskConsumer interface. The latest features are in the ISASTaskConsumer3 interface, which provides more access to application options and project information. Existing tasks that use the ISASTaskConsumer interface will work just fine, and they do not need to rebuild or change at all.

Likewise, the ISASTask3 interface provides a more elegant method for tasks to communicate information about output data. (ISASTask3 uses a new interface named ISASTaskDataDescriptor.) If an existing task uses the older ISASTask instead, the application still supports it. However, new tasks that use the newer version of the interface can take advantage of any new methods or approaches.

Understanding the Life Cycles of Your Task

In software, the term *life cycle* means several things, including the methodology of how you approach the software development process. But, in this chapter, *life cycle* has a much more practical meaning. Simply, it's the series of events that occur when your task is installed, opened and shown, and run.

Each of these life cycle phases accesses a different set of properties and methods and usually in a predictable sequence. Here are the three phases:

1. The task is installed and "discovered" by the application. The task name and icon appear in the application's menu.
2. The task is opened and shown (such as when a user selects the task from a menu).
3. The task is run. The term run means that the task is called to generate a SAS program or do work that creates results.

Each of these phases represents a different life cycle for the task. Your implementation of the custom task APIs is usually encapsulated in a single .NET class. Think of the task's life cycle as the sequence of property and method calls that occur within this single .NET class, from the time the class is constructed (using its default .NET constructor) until the time the class is released.

Life Cycle of a Task during Discovery

The discovery phase of a task occurs when you add a task to the application's menu using one of the following mechanisms:

- You select **Tools→Add-In→Add-In Manager** in SAS Enterprise Guide, or you run regaddin.exe in the SAS Add-In for Microsoft Office.
- The application adds it. The task assembly (which is a DLL file) is in a special folder that the application searches for and opens at start-up. See the section in Chapter 1 about "Deploying Custom Tasks" for more information about special folders.

The purpose of the discovery phase is to gather enough information about the task so that the application can categorize it and place it in a menu. It's a good idea to make sure that the supporting methods and properties are lightweight. They should require very little processing and a minimum of dependencies. If the methods take a long time to execute, the user will experience delays simply trying to open the application.

When a task is being discovered, the following methods and properties are called:

- Task class constructor.
- ISASTaskDescription.Languages, which gets a list of the languages (or cultures) that are supported by the task.
- For each supported language, ISASTaskDescription.Language(set), which sets the current language. In addition, ISASTaskDescription.FriendlyName, ISASTaskDescription.WhatIsDescription, and ISASTaskDescription.TaskName, which are localizable values.
- The remaining culture-neutral properties of ISASTaskDescription.

Life Cycle of a Task during Open and Show

The open and show phase involves the path that is followed when a user selects the task from a menu or selects an existing instance of the task and then selects **Modify**. (**Modify** is the name of the menu option in versions 4.2 and 4.3 of SAS Enterprise Guide and the SAS Add-In for Microsoft Office. In earlier versions, the menu option is named **Open**.)

When a task is opened or modified from an existing instance, the following methods and properties are called:

- Task class constructor.
- ISASTaskAddIn.Connect(), which connects the application to the task. The Connect() call passes a reference to ISASTaskConsumer, which is a handle to the application. During

Connect() processing, you can gather information about the input data sources and any application options in your task.

- ISASTask.Initialize(), which provides a good place to initialize any data structures or objects that your task might need. It's better to place initialization logic in this location rather than in the task class constructor because you don't need initialization overhead during the discovery phase.

- ISASTask.XmlState(set), which provides the task state information that was saved. When the user modifies an existing task in a project, this method passes the existing task state so that you can initialize your task structures and user interface to show the current settings.

- ISASTask.Show(), which is your cue to show the task's user interface. Usually, you provide a modal window as a user interface, which allows the user to interact with the task but not with any other part of the application. A user usually completes the task by saving changes (clicking **Finish**, **Run**, or **OK**) or by discarding changes (clicking **Cancel**). With the Show() method, a ShowResult value is returned and indicates the next action—RunNow, RunLater, or Canceled. If the ShowResult.Canceled value is returned, then you skip directly to the cleanup routines (Disconnect and Terminate).

- ISASTask.XmlState(get), which captures the new and revised task state information for use in the project. With the task settings captured, the application can safely close this instance of the task.

- ISASTask.Terminate(). Not many tasks have anything to do with this method because .NET automatically "cleans up" objects when they are no longer in use. (This happens via the .NET "garbage collection" mechanism.) But, if your task has references to any unmanaged resources (such as data connections), this method closes and releases them.

- ISASTaskAddIn.Disconnect(), which is where the application formally says goodbye to this instance of the task. At this point, you should not make any further calls to the application using the ISASTaskConsumer interface. The application has effectively hung up the phone. If you saved a reference to ISASTaskConsumer, you should set it to Null (or Nothing in Visual Basic).

After the Disconnect() method is called, the remaining life of your task is controlled by .NET memory management. The .NET garbage-collection routines clean up and free memory when appropriate.

Life Cycle of a Task during Run

The run phase occurs when a user runs a project or flow that contains an instance of the task. Or, it occurs when a user selects an instance of the task and selects **Refresh** (or **Run** in SAS Enterprise Guide 4.1). This phase is very similar to the open and show phase. Instead of calling ISASTask.Show() to show the task's user interface, the run phase calls the methods that enable the task to create results by returning a SAS program or performing an operation.

When a task is run, the following methods and properties are called:

- Task class constructor.
- ISASTaskAddIn.Connect(), which connects the application to the task.
- ISASTask.Initialize(), which initializes any data structures or objects.
- ISASTask.XmlState(set), which provides the task state information that was saved.
- ISASTask.SasCode(get), which generates a SAS program that you want to run at this point in the project. Ideally, the process of generating a SAS program to run should be quick. Remember, this step simply generates the SAS program to run. It does not actually run the program. The application (such as SAS Enterprise Guide) submits the program to run in a SAS session after this instance of the task has been disconnected and terminated.
- ISASTask.OutputDataCount(get) and ISASTask.OutputDataInfo(get), which gather information about the output data sets that you expect your task to create. This information can help the application populate the results with the appropriate output data set references when your SAS program is run. As an alternative in version 4.2 and later in SAS Enterprise Guide, you can use ISASTask3.OutputDataDescriptorList, which is a more elegant method.
- ISASTask.Terminate(), which closes and releases any objects.
- ISASTaskAddIn.Disconnect(), which is where the application says goodbye to this instance of the task.
- During the run phase, the task class is created for only one purpose—to generate the SAS program that you want the application to run based on the task settings and the application environment (such as the input data source and application options).

Exercise: Observe the Life Cycle of a Task

To better understand a task's life cycle, it might be helpful to see what happens behind the scenes. To facilitate this, I've created a bare-bones custom task that creates log messages for each stage in the cycle, from initialization to running to serialization to termination.

The log messages are created using log4net, a technique that is described in greater detail in Chapter 8, "Debugging Techniques: Yes, You Will Need Them."

Here's how to observe the task and review its activities:

1. Download the example custom task, SASPress.Examples.Lifecycle.dll, from the SAS author page for this book, http://support.sas.com/hemedinger.
2. Install the task by copying it to the appropriate `Custom` folder based on the version of SAS Enterprise Guide that you are using.
3. Copy the logging.config file, which is provided with the SASPress.Examples.Lifecycle.dll task, to your user profile area. Your user profile area has the same root path as the `Custom` folder (for example, `%appdata%\SAS\EnterpriseGuide\4.3`).

4. Open SAS Enterprise Guide. Select **Tools→Add-In** and verify that the **Log Task API calls** task appears. But, do not select it, yet! Without doing anything, close SAS Enterprise Guide.
5. There should be a `Logs` folder in your user profile area. The file with the most recent timestamp contains the sequence of events that occur when SAS Enterprise Guide discovers a task and adds it to a menu.

In the discovery phase, the log of the sequence of events looks something like this partial output:

```
API: In task constructor
API: ISASTaskAddIn.Languages(out string[] items)
API: ISASTaskAddIn.Language - set(en)
API: ISASTaskAddIn.AddInDescription - get()
API: ISASTaskAddIn.AddInDescription - get()
API: ISASTaskDescription.TaskName - get()
API: ISASTaskDescription.TaskDescription - get()
API: ISASTaskDescription.TaskCategory - get()
API: ISASTaskDescription.WhatIsDescription - get()
API: ISASTaskDescription.Validation - get()
API: ISASTaskDescription.FriendlyName - get()
API: ISASTaskAddIn.Language - set(en-US)
API: ISASTaskAddIn.AddInDescription - get()
```

Each logged event is prefixed by the task's "logger" name:

```
SASPress.Examples.Lifecycle.Logevents
```

Here's how to observe the sequence of events when the task was in the open and show phase:

1. Open SAS Enterprise Guide.
2. Select **Tools→Add-In→Log Task API calls**. The Log Task API calls dialog box displays, as shown in Figure 4.1.

Figure 4.1: The Log Task API Calls Dialog Box

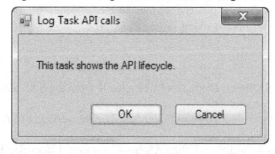

3. Click **Cancel** to close the window. Close SAS Enterprise Guide.
4. Go back to the `Logs` folder. There should be a new log file with a more recent timestamp.

The content of the newer log file shows the discovery phase activity, just like it did in a previous test. But, this time, there is another block of log entries that includes the events that occur when you invoke the task user interface. This block includes user interface-related methods, such as:

```
API: ISASTask.Label - get()
API: ISASTask.Show(IWin32Window Owner=SEGuide.exe)
API: in Show(), showing task form
API: leaving Show(), task form closed with Cancel action
```

To observe the events that occur when you run your task, repeat the previous steps, but, this time, go ahead and run the task.

The **Log Task API calls** task generates a simple SAS program that runs in your SAS session. After the task completes, close SAS Enterprise Guide. Once again, go back to the **Logs** folder to find the log file with the most recent timestamp.

The content of the latest log file shows the activities of all three phases—discovery, open and show, and run. The run phase contains entries such as:

```
API: ISASTask.Initialize()
API: ISASTask.XmlState - set()
API: ISASTask.Label - set('Log Task API calls')
API: ISASTask.OutputDataCount - get()
API: ISASTask.OutputDataInfo(int Index=0,
    out string source, out string label
API: ISASTask.OutputDataCount - get()
API: ISASTask.Label - get()
API: ISASTask.SasCode - get()
```

This log output shows how the task is created and then initialized with its saved XML state. In the last line, SAS Enterprise Guide asks the task for the SAS program to run.

Special Interfaces for Special Tasks

The methods described in the previous section apply to the typical SAS task—a task that has the job of creating a SAS program to run. However, what if you don't want to generate a SAS program? What if you want a task that performs a different operation that involves something different from a SAS program or in addition to a SAS program?

Fortunately, the custom task APIs provide hooks for special tasks. You can use ISASTaskExecution to provide greater control over how your task works. In SAS Enterprise Guide, additional interfaces, such as ISASProject and ISASProjectProcessFlow, programmatically explore and access the content of your current project.

Greater Control Using ISASTaskExecution

The ISASTaskExecution method, which you can use in your custom task, puts you in the driver's seat when it comes to running your task.

To better understand the need for this control, consider a built-in task in SAS Enterprise Guide. Beginning in release 4.2, SAS Enterprise Guide includes a task in the **Data** category called **Upload Data Files to Server**. This task enables you to select SAS data files from your local computer and upload them to your SAS server. Once the data files are in SAS, you can perform additional analyses, run programs with them, or run queries or tasks with them. (There is a similar task in this category that does the reverse called **Download Data Files to PC**. This task downloads SAS data files from the SAS server to your local PC.)

Some of this work falls outside of what you can do with a SAS program. The **Upload Data Files to Server** task works by using a custom API on the SAS workspace connection. This custom API copies file contents from your local PC to the SAS server. To finish the job, the task runs a short SAS program to ensure that the data files are in the proper encoding and format for the target SAS server.

If you examine the log output from the **Upload Data Files to Server** task, you'll see a mixture of custom output and SAS program log statements. In this example log output from the task, the lines that are prefixed with numbers are part of the SAS program log. The other lines are informative statements created by the task.

```
Copying SAS data files to Local
Destination SAS library is: WORK
Copied "C:\SAS\Data\Sample\bandaid.sas7bdat" to
    "C:\tmp\bandaid.sas7bdat"
Copied 5,120 bytes in 0.11 seconds.
Data file representation (WINDOWS_32 wlatin1  Western
(Windows)) does not match target system.
Rewriting data file.  SAS log:
1          ;*';*";*/;quit;run;
2          OPTIONS PAGENO=MIN;
3          DATA WORK.bandaid;
4            SET WORK.bandaid;
NOTE: Data file WORK.BANDAID.DATA is in a format that
is native to another host, or the file encoding does
not match the session encoding. Cross Environment Data
Access will be used, which might require additional CPU
resources and might reduce performance.
5          RUN;

NOTE: There were 4 observations read from the data set
WORK.BANDAID.
NOTE: The data set WORK.BANDAID has 4 observations and
3 variables.
```

```
NOTE: DATA statement used (Total process time):
      real time              0.01 seconds
      cpu time               0.00 seconds
6
7         QUIT; RUN;
8
Total transfer: 5,120 bytes in 0.50 seconds
Total number of files processed: 1
```

For an example of how to use ISASTaskExecution in a task, see Chapter 12, "Abracadabra: Turn Your Data into a SAS Program," and Chapter 14, "Take Command with System Commands."

Access the Project Contents with ISASProject

To indirectly borrow from a famous poem by John Donne, no task is an island. In the early days of custom tasks, developers yearned for a way to access the contents of a SAS Enterprise Guide project from within the project itself. For example, suppose you want your custom task to see what other tasks are in the process flow, and then to determine whether the task complies with a corporate or group standard.

In SAS Enterprise Guide 4.2, APIs were added to make project access within a project possible.

Note: These APIs work only in SAS Enterprise Guide, not in the SAS Add-In for Microsoft Office. Creative developers in the SAS Add-In for Microsoft Office can already navigate through built-in object models for Microsoft Office documents without the need for any special APIs.

Here are the APIs:

ISASProject
Provides access to basic project information, such as the project name and the date it was created. The most important role of ISASProject is to provide access to the project's process flows. Process flows are accessed using ISASProjectProcessFlow. The ProcessFlows property provides a list of ISASProjectProcessFlow interface handles, one for each process flow in the project. (Every project has at least one process flow.)

ISASProjectProcessFlow
Provides access to the contents of a process flow. Using this API, you can get tasks (using the GetTasks() call) and perform interesting operations on the items in the process flow, such as check whether items are linked, add your own links, and gather the SAS code that would be generated by the entire flow or just a subset of tasks in the flow.

ISASProjectTask
Represents an individual task in a process flow. Using this API, you can get a list of input data (if there is any) for the task and the task type and its internal state (the guts of the XmlState property). The task object that you access implements ISASProjectItem, which provides basic information such as the dates the task was created and modified, the name, and even the icon that represents the task.

The ISASTaskConsumer3 interface provides access to the project-related APIs. Depending on which part of the project you want to work with, you can use one of these methods:

- GetProject(), which provides the ISASProject interface for the current project.
- GetThisProcessFlow(), which provides the ISASProjectProcessFlow interface for the active process flow. (The active process flow is where your active task currently sits.)
- GetThisTask(), which provides the ISASProjectTask interface for the task that you're currently working with.

For a real-life example of the ISASProject interfaces in action, see Chapter 10, "For the Workbench: A SAS Task Property Viewer."

More Interfaces That Play Nice

Users of SAS Enterprise Guide and the SAS Add-In for Microsoft Office often have an expectation that tasks behave in a consistent way. This expectation is regardless of who developed the tasks or whether they are built-in or supplied by a third party. (This expectation is yet another way that users like to keep us developers on our toes.)

Many of the built-in tasks have valuable behaviors that you might want to emulate in your own tasks.

Most built-in tasks:

- Respect the application options (which are usually set in **Tools→Options**) that control default output settings and other common preferences.
- Support the use of task templates, which enable a user to save task settings as a "favorite" for later use.
- Respect the efforts of an administrator to restrict certain application behaviors, such as whether to allow access to types of data or files or whether to allow a specific task to run at all. (These restrictions are set in SAS Management Console, where an administrator can control which application capabilities are allowed for users with defined roles.)

To set these behaviors in your tasks, use these APIs:

ISASApplicationOptions
   Provides easy access to basic application options that you might need to know about, such as the preferred location for output data and the preferences for ODS output from SAS. You can get this interface by calling ISASTaskConsumer3.GetApplicationOptions().

ISASTaskTemplate
   Tells the SAS application whether your task supports task templates, which the application can manage for you. A task template is a saved collection of task settings. A task template enables

the user to later use the task settings in another project or with a different data source. To learn more about task templates, see the SAS Enterprise Guide online Help. This API is implemented in the custom task.

ISupportRoleBased
Enables your task to be a good citizen if an administrator has restricted certain application behaviors. If you implement ISupportRoleBased, then the task can provide information about restrictions that you should follow, such as whether your task should be Read-Only or whether the user can redirect output data that your task creates.

Chapter Summary

We've talked about a lot of different interfaces, each with its own properties and methods to do the work that custom tasks need to do. In the later chapters that contain specific examples, you'll see these interfaces exercised to achieve the desired results. Some interfaces are ubiquitous—every task must use them to plug in to SAS Enterprise Guide. Other interfaces have very specific purposes—only certain tasks need to use them. As a developer of custom tasks, the more familiar you are with the interfaces and APIs, the more successful you'll be when you introduce your task into the ecosystem of a SAS application.

Chapter 5: Meet the Task Toolkit

Task Toolkit: What's In It? **55**
    The SasTask Class 56
    The SasServer Class and SAS.Tasks.Toolkit.Data Namespace 59
    The SAS.Tasks.Toolkit.SasSubmitter Class 60
    The SAS.Tasks.Toolkit.Helpers Namespace 61
    The SAS.Tasks.Toolkit.Controls Namespace 62
    Examples of Using the SAS.Tasks.Toolkit Classes 63
Chapter Summary **67**

You met the custom task APIs in Chapter 4. They are powerful and flexible, but you might have found them to be a bit overwhelming to navigate. After all, the APIs do represent a contract, and contracts can be complicated (and when broken, they can be unforgiving).

Because not all people who want to create custom tasks are versed in the nuances of an application contract, SAS provides a library of classes that make it much easier to comply with a contract. This library is called the Task Toolkit.

This chapter describes the main features of the Task Toolkit. By using the classes in the Task Toolkit, you end up maintaining much less of the plumbing code required by a custom task. Instead, you can devote your energy to the functional and business purposes of your task—the stuff that makes the task valuable for you and your users.

Task Toolkit: What's In It?

Just as the custom task APIs have their own namespace (`SAS.Shared.AddIns`) and their own .NET assembly (SAS.Shared.AddIns.dll), the Task Toolkit classes have their own namespace, `SAS.Tasks.Toolkit`. The .NET assembly is SAS.Tasks.Toolkit.dll.

Note: The Task Toolkit library was added in SAS Enterprise Guide 4.2. (It continues to be available in version 4.3 and later.) SAS does not provide a similar toolkit for SAS Enterprise Guide 4.1.

Here is a high-level overview of what the Task Toolkit does for you:

- Makes it easy to implement the custom task APIs. The `SAS.Tasks.Toolkit.SasTask` class implements most of what you need from `ISASTask`, `ISASTaskDescription`, and `ISASTaskAddIn`. Hundreds of lines of code can be reduced to a handful.

- Provides easy access to objects related to SAS. The `SAS.Tasks.Toolkit.SasServer` class provides a useful object representation of your SAS Workspace Server. Several classes in the `SAS.Tasks.Toolkit.Data` namespace enable you to access SAS data, libraries, columns, and even SAS catalogs.

- Provides access to SAS services such as submitting SAS programs. Classes and methods in the Task Toolkit make it easy to submit SAS programs, synchronously or asynchronously. You can retrieve the SAS log and collect results from the programs.

- Provides a library of useful user interface controls. The Task Toolkit includes common controls that tasks can use to navigate data sources, display SAS programs, and more. (These controls are implemented using Windows Forms in SAS Enterprise Guide 4.2, 4.3, and 5.1.)

The SasTask Class

The `SAS.Tasks.Toolkit.SasTask` class takes most of the tedium out of implementing a custom task. The `SasTask` class contains default implementations for dozens of required methods and properties. If you examine the `SasTask` class definition in the Object Browser in Visual Studio, you will see that it implements at least 10 interfaces! Here is the class declaration for the `SasTask` class in SAS Enterprise Guide 4.2:

```
public class SAS.Tasks.Toolkit.SasTask :
  System.ComponentModel.Component,
  SAS.Shared.AddIns.ISASTaskAddIn,
  SAS.Shared.AddIns.ISASTaskDescription,
  SAS.Shared.AddIns.ISASTask3,
  SAS.Shared.AddIns.ISASTask2,
  SAS.Shared.AddIns.ISASTask,
  SAS.Shared.AddIns.ISASTaskSubmitSink,
  SAS.Shared.AddIns.ISASTaskDataSink2,
  SAS.Shared.AddIns.ISASTaskDataSink,
  SAS.Shared.AddIns.ISASTaskTemplate,
  SAS.Shared.AddIns.ISupportRoleBased;
```

All of these interfaces (which are often repetitive and not always pertinent to your needs) are handled with an easy-to-use base class. When you have your custom task inherit from `SasTask`, the amount of code in your custom task is drastically reduced.

Here is a C# language example of a valid custom task implementation that inherits from SasTask:

```csharp
// unique identifier for this task
[ClassId("a3cd0efb-062a-47b2-8bf9-777ec3c530f0")]
[InputRequired(InputResourceType.None)]
public class TaskExampleTask : SAS.Tasks.Toolkit.SasTask
{
  public TaskExampleTask()
  {
    InitializeComponent();
  }
  private void InitializeComponent()
  {
    this.TaskCategory = "SAS Custom Task";
    this.TaskDescription = "Task Description";
    this.TaskName = "TaskExampleTask";
  }
}
```

That's it. Technically, this is a valid custom task implementation. But, even though it is valid, it's useless as is. The task doesn't have methods for showing the user interface or generating a SAS program or saving and restoring the task state.

It's easy to add that logic into the mix. You simply override the methods that your task requires. For example, to show the user interface, override the `Show()` method with the code to show a form or window. To generate a SAS program, override the `GetSasCode()` method. And, to save and restore the task state, override `GetXmlState()` and `RestoreStateFromXml()`, respectively.

Here is a more complete example of a class that inherits SasTask and overrides the methods in place. The example uses Visual Basic as the programming language. (An underscore at the end of a line signifies the line-continuation character in Visual Basic. This example is formatted using line-continuation characters so that it can fit correctly within the margins of this book.)

```vb
<ClassId("a5370964-0383-46a7-91a4-84afd437f54c")> _
<IconLocation("ExampleTask.task.ico")> _
<InputRequired(InputResourceType.Data)> _
Public Class ExampleTask : _
  Inherits SAS.Tasks.Toolkit.SasTask
  Sub New()
    InitializeComponent()
  End Sub
  Sub InitializeComponent()
    Me.ProductsRequired = "BASE"
    Me.TaskCategory = "SAS Custom Task"
    Me.TaskDescription = "Task Description"
    Me.TaskName = "ExampleTask"
  End Sub
  Private _settings _
    As New ExampleTaskSettings
```

```vbnet
' This function is called when it's
' time to show the task window
Public Overrides Function _
   Show(ByVal Owner As IWin32Window) _
As SAS.Shared.AddIns.ShowResult
   Dim form As New ExampleTaskForm
   ' initialize with the Consumer member
   ' and Settings values
   form.Consumer = Consumer
   form.Settings = _settings
   ' if OK, then capture the settings values
   If (DialogResult.OK = _
   form.ShowDialog(Owner)) Then
      _settings = form.Settings
      Return ShowResult.RunNow
   End If
   Return ShowResult.Canceled
End Function
' This function is called when it's
' time to generate a SAS program
' from the task settings so far.
Public Overrides Function GetSasCode() _
   As String
   Dim code As String
   ' generate the SAS program, based on
   ' current settings, and return as a String
   code = String.Format("proc options option={0}; run;", _
         _settings.SasOption)
   Return code
End Function
Public Overrides Function GetXmlState() _
   As String
   Using sw As New StringWriter()
      Dim s As New XmlSerializer( _
      GetType(ExampleTaskSettings)_
      )
      s.Serialize(sw, _settings)
      Return sw.ToString()
   End Using
End Function
Public Overrides Sub RestoreStateFromXml( _
   ByVal xmlState As String)
   Using sr As New StringReader(xmlState)
      Dim s As New XmlSerializer( _
      GetType(ExampleTaskSettings)_
      )
      _settings = CType(s.Deserialize(sr), _
            ExampleTaskSettings)
   End Using
End Sub
End Class
```

This example relies on other classes as well. For example, there is the ExampleTaskSettings class that encapsulates the logic to convert the task state to and from XML. The ExampleTaskForm class is the Windows Form that is displayed when you start or modify the task. These two classes require additional classes that you must create to do the work of your task. Keep in mind, though, that SasTask takes care of the plumbing, so you can concentrate on what gives your task real value (the user interface, the SAS program that it generates, and the business logic that's needed to glue it all together).

The SasServer Class and SAS.Tasks.Toolkit.Data Namespace

Almost every custom task needs to have access to one or more data sources. The Task Toolkit provides classes that help you navigate the data sources that you can access from your SAS environment.

Here is a list of the most useful classes related to accessing data sources:

SAS.Tasks.Toolkit.SasServer

Represents a SAS Workspace Server. Even though this class is not strictly part of the Data namespace, you can use `SasServer` to get a list of SAS libraries, which, in turn, enables you to explore the data sources that are available in the SAS session. To get the list of SAS libraries, use the `GetSasLibraries` method.

Because you might have access to multiple SAS servers in SAS Enterprise Guide, the `SasServer` class provides the `GetSasServers` method to get a list of all SAS servers that are available to you.

SAS.Tasks.Toolkit.Data.SasLibrary

Represents a SAS library. There are two methods available in this class that give you access to data objects. `GetSasDataMembers` gives you access to data set and data view members. `GetSasCatalogMembers` gives you access to SAS catalogs, which contain SAS catalog entries. The `SasLibrary` class holds properties for the library, such as the library engine (for example, BASE or ORACLE), the path, whether it's Read-only (assigned ACCESS=READ), and additional options.

SAS.Tasks.Toolkit.Data.SasData

Represents a data set or data view. This might be a native SAS data set if the SAS library is a Base SAS library. Or, it can be a database table, such as a Microsoft Access table or Oracle table if the library was assigned using a SAS/ACCESS library engine. The `SasData` class has several useful methods and properties. The most useful method is `GetSasColumns`, which returns a list of SAS columns or variables that reveal what types of data are in the data source.

SAS.Tasks.Toolkit.Data.SasColumn

Represents a column or variable in a data set or data view. Each `SasColumn` object provides access to the standard attributes that you expect from a SAS data variable, including name, label, type (numeric or character), format, informat, length, and more.

SAS.Tasks.Toolkit.Data.SasCatalog and SAS.Tasks.Toolkit.Data.SasCatalogEntry
SAS catalogs can reside in SAS libraries. They work similar to folders in a file system. SAS catalogs contain catalog entries of various types and are used most often by legacy SAS applications. However, you might find them useful to store and retrieve information in a platform-independent way. (For an example of a task that uses these classes, see Chapter 16, "Building a SAS Catalog Explorer.")

These classes are designed so that you can approach your data from an exploratory point of view. You can access a list of servers and select one, and then you can access the libraries from that server. You can select a library, and then you can access a list of data members. When you select a data source from that list, you can obtain the list of data columns for that data source.

The SAS.Tasks.Toolkit.SasSubmitter Class

The SasSubmitter class provides simple methods for submitting sections of SAS code for processing, and simple methods for retrieving the SAS log output from the programs that you submit.

You initialize the SasSubmitter class with the name of the SAS server that you want to process your program. For example, in C#:

```
using SAS.Tasks.Toolkit;
SasSubmitter submitter =
   new SasSubmitter(Consumer.AssignedServer);
```

Here are some of the most useful methods:

IsServerBusy
Checks to see whether the selected SAS server is busy processing another program. A SAS Workspace Server can process only one program at a time. If the user is doing other work in SAS Enterprise Guide, it is possible that the SAS server is occupied. (In SAS Enterprise Guide, each user gets his or her own SAS session. You don't have to worry about contending with other users in the same SAS environment.)

You can submit another program even when the SAS Workspace Server is currently busy. SAS Enterprise Guide adds the program to a queue to wait for processing. If your goal is to provide a near-instant response to the user, you should check whether the server is busy and, if so, provide feedback to the user to try again later.

SubmitSASProgramAndWait
Provides the simplest model for running a SAS program. It provides *synchronous* execution, which means that the method does not return a value until the program has completed processing. The method returns a Boolean value indicating success. If the SAS program generates an error, the return code is **false**.

The `SubmitSASProgramAndWait` method provides the SAS log as an output parameter. You can use .NET string methods or regular expressions to parse the log for certain content. For example, if you want to check the log for errors or warnings, you can parse the log.

Note: The wait behavior of this method can cause problems if the program fails to complete for any reason. For the most robust behavior, use the asynchronous `SubmitSASProgram` method instead, even though it involves more coding logic.

SubmitSASProgram
Submits a SAS program using an *asynchronous* model, which means the control returns to your task code as soon as you call the method. To track the progress of the program and retrieve the results, you must add a handler for the `SubmitSASProgramComplete` event.

The advantage of this model is that your task appears much more responsive. If the user clicks a button that runs some SAS code, the task's user interface won't be blocked while the program is running. However, the `SubmitSASProgramComplete` event adds complexity to your code.

Note: The SAS.Tasks.Toolkit.SasSubmitter class is available in SAS Enterprise Guide 4.3 and later. If you have SAS Enterprise Guide 4.2, you can use the `SubmitSASProgram` and `SubmitSASProgramAndWait` methods in the SasTask class instead.

The SAS.Tasks.Toolkit.Helpers Namespace

As you create custom tasks, you'll find that there are many recurring patterns and routines that you need to ensure good quality. Attending to these details can be tedious, but it's necessary to create robust tasks that don't fall apart at the first sign of a nonstandard variable name or some other edge condition.

Before you invest time in creating your own routines to perform common operations, check out what the SAS.Tasks.Toolkit.Helpers namespace has to offer. It contains several helpful classes that answer needs that are common to many custom tasks.

Here are some of the especially useful classes:

SAS.Tasks.Toolkit.Helpers.TaskDataHelpers
Makes it easier to access data, which is fundamental to many custom tasks. For example, this class contains a method named `GetDistinctValues()`, which enables you to retrieve a list of distinct values from a column in a data set. This is very useful for populating a user control with choices that are fed from a data set.

This class contains a method named `GetSasCodeReference()`, which generates a valid SAS data reference for use in a SAS program. For example, if the active data is named Analysis and is in the Work library, this method returns the value Work.Analysis. That seems pretty simple, but the method also recognizes other things such as password-protected data or observation limits that the user has specified. The method returns these as data set options in the SAS code reference. As a result, this method removes some of the fuss around having to generate this syntax yourself.

SAS.Tasks.Toolkit.Helpers.TaskAddInHelpers
Makes contracts easier to implement. One particularly useful method is `CreateTaskTemplate()`, which enables your task to easily integrate with the Task Template feature of SAS Enterprise Guide 4.2 and 4.3. Task templates enable a user to store favorite task settings in his or her environment, and then call them up later in other projects. Users can easily share their favorite task settings with other users.

SAS.Tasks.Toolkit.Helpers.TaskOptions
Enables users to override behaviors. SAS Enterprise Guide users can override specific behaviors of a task by modifying options or task properties. Options include settings such as which SAS library to use as a default output location, what types of ODS results to create (HTML, RTF, SAS Report, and so on), the default footnote or title text for a report, and more. The `TaskOptions` class offers properties and methods that enable you to access and change the values of these options so that your custom task can comply with the user's preferences.

For example, you can call `GetDefaultFootnoteText()` to get the preferred text for the `FOOTNOTE` statement. Most tasks allow the user to override the footnote text, but by seeding your task with a good default value for the footnote, you might save your users time.

SAS.Tasks.Toolkit.Helpers.UtilityFunctions
Helps you avoid lots of little gotchas that you might encounter when you programmatically generate SAS programs.

For example, SAS programmers know that a SAS variable name must conform to certain rules. The traditional rules assume that a variable name is alphanumeric, contains no spaces or special characters, and is 32 characters or less. However, it is possible in SAS to stray from those rules. But, when you do, you must use special syntax called a SAS name literal. Use the `SASValidLiteral()` function to ensure that a variable name that you format in a SAS program uses the correct syntax.

The `GetValidSasName() function` enables you to specify any string and create a valid name for use in SAS syntax. A valid name is alphanumeric, contains no spaces or special characters, and is 32 characters or less.

The SAS.Tasks.Toolkit.Controls Namespace

A clean and consistent user interface is important for a custom task. Because you might want your task's user interface to be consistent with SAS user interfaces, the SAS.Tasks.Toolkit.Controls namespace offers classes that enable you to leverage some of the built-in user interface elements from SAS.

These elements don't offer everything that you need to provide that SAS look and feel, but they do offer an easy way to provide some of the complex functions that are in many SAS tasks.

SAS.Tasks.Toolkit.Controls.SASTextEditorCtl

The SAS Program Editor appears prominently in SAS Enterprise Guide. It allows you to display and edit SAS programs, and provides useful features such as keyword color-coding and syntax completion. It's an impressive editor with tremendous capabilities.

With the SASTextEditorCtl class, you can include the Program Editor so that users can view and edit SAS program statements in your own task. It's easy to include with just a few code statements.

You can use the SASTextEditorCtl control in almost the same way you use a TextBox control. Here are the C# statements:

```
using SAS.Tasks.Toolkit.Controls;
// declare in your Form class
private SASTextEditorCtl ctlEditor;
private SASTextEditorCtl ctlLogViewer;
ctlEditor = new SASTextEditorCtl();
ctlLogViewer = new SASTextEditorCtl();
ctlEditor.ContentType =
   SASTextEditorCtl.eContentType.SASProgram;
ctlLogViewer.ContentType =
   SASTextEditorCtl.eContentType.SASLog;
// add to the current form
this.Controls.Add(ctlEditor);
this.Controls.Add(ctlLogViewer);
```

For a complete example that shows how to include the SAS Program Editor in a custom task, see Chapter 12, " Abracadabra: Turn Your Data into a SAS Program".

Simpler to use but a little less flexible are two helper classes that display SAS program code and SAS log content in a modal dialog box. These two helper classes are `SASCodeViewDialog` and `SASLogViewDialog`, respectively. An example of the `SASLogViewDialog` is provided in the next section.

Examples of Using the SAS.Tasks.Toolkit Classes

This section provides examples of some of the most commonly used SAS.Tasks.Toolkit classes and methods.

Example: Describe the Scope and Behavior of a Task

When you define a custom task using the SasTask class, you can associate different attributes with the task that affect how SAS Enterprise Guide treats it. For example, you can assign a task category name so that similar tasks are grouped together in the application's menu. You can tell SAS Enterprise Guide that your task requires input data or that your task should be available only if it is connected to a SAS metadata environment.

Many of these are simple to code by using class-level attributes—the bracketed keyword constructs that precede the class definition.

The following example tells SAS Enterprise Guide that input data is required. As a result, SAS Enterprise Guide prompts for a data source if one is not yet selected. The code specifies that SAS metadata is required. If the user doesn't have an active SAS metadata profile, then the task appears as disabled in the menu. This enables you to use role-based capabilities (described in Chapter 1, "Why Custom Tasks") to prevent unauthorized users from running the task.

```
// unique identifier for this task
[ClassId("ab4f1d58-6c07-4c6b-abab-43525e6a8a3f")]
// location of the task icon to show in the menu
[IconLocation("Controlled.task.ico")]
[InputRequired(InputResourceType.Data)]
[SASMetadataRequired(true)]
[Version(4.3)]
public class ControlledTask :
    SAS.Tasks.Toolkit.SasTask
{
// task implementation follows
}
```

Example: Navigate the SAS Server and Library Hierarchy

The following example, written in C#, shows how you can start with a list of SAS servers, narrow your selections down to a single column, and display just the properties of the selected column. The code example is incomplete. It shows only the methods that you use in the Task Toolkit to retrieve the lists of the various objects. The comments in the code show the places where you have additional logic to narrow down a list of objects to a single selection.

```
using SAS.Tasks.Toolkit;
using SAS.Tasks.Toolkit.Data;
public void Explore()
{
   SasServer selectedServer = null;
   SasLibrary selectedLibrary = null;
   SasData selectedMember = null;
   SasColumn selectedColumn = null;
   List<SasServer> servers = SasServer.GetSasServers();

   /* ... code to select a server */
   ReadOnlyCollection<SasLibrary> libraries =
     selectedServer.GetSasLibraries() as
       ReadOnlyCollection<SasLibrary>;
   /* ... code to select a library */
   ReadOnlyCollection<SasData> members =
     selectedLibrary.GetSasDataMembers() as
       ReadOnlyCollection<SasData>;
   /* ... code to select a data member */
```

```
    ReadOnlyCollection<SasColumn> columns =
      selectedMember.GetColumns() as
        ReadOnlyCollection<SasColumn>;
   /* ... code to select a column */

    string properties =
      string.Format("Name: {0}, Type: {1}, Format: {2}",
        selectedColumn.Name,
        selectedColumn.Type.ToString(),
        selectedColumn.Format);
 }
```

If you already know the exact data member that you need to access, the Task Toolkit classes provide convenient constructors to create the necessary objects with just a few lines of code. For example, if you want to get a list of the columns in the Sashelp.Cars data set from the SAS server named SASApp, the code would look something like this:

```
 SasData data = new SasData("SASApp", "SASHELP", "CARS");
 ReadOnlyCollection<SasColumn> columns =
   data.GetColumns() as
   ReadOnlyCollection<SasColumn>;
```

Here is the same example using Visual Basic:

```
 Dim data As New SasData("SASApp", "SASHELP", "CARS")
 Dim columns As _
   ReadOnlyCollection(Of SasColumn) = _
   data.GetColumns()
```

The `SasServer` and `SasLibrary` classes offer similar shortcut constructors to help you quickly access the objects that you need.

Example: Check Whether a SAS Library Connects to a Database

The SasLibrary class includes a static method to check whether a specified SAS library is considered a DBMS. This is important if your task generates SQL statements and you want to make sure that those statements will pass through to the database for processing.

```
 using SAS.Tasks.Toolkit.Data;
 /* libref is a string that contains the syntax name */
 /* of the SAS library */
 SasLibrary lib =
   new SasLibrary(Consumer.AssignedServer, libref);
 bool isDbms = false;
 string engine = lib.Engine;
 isDbms = SasLibrary.IsDBMS(engine);
```

In the output, the value of `isDbms` will be false for SAS file-based libraries that use the BASE engine. The value will be true for DBMS library engines supported by SAS/ACCESS, such as ORACLE or ODBC.

Example: Retrieving the Value of a SAS Macro Variable

The following C# statements use the SasServer class to connect to a SAS Workspace Server and detect the version of SAS that's running (using the built-in SYSVER macro variable).

```
using SAS.Tasks.Toolkit;
/* ... */

/* Consumer.AssignedServer is the active server from SAS EG */
SasServer s = new SasServer(Consumer.AssignedServer);
double dServerVer = 9.2;
string ver = s.GetSasMacroValue("SYSVER");
try
{
    dServerVer = Convert.ToDouble(ver);
}
catch
{
/* Convert can throw exception, but should not in this case */
}
```

In the output, the value of `dServerVer` should be 9.2 or 9.3, depending on the version of SAS that is used. By converting the value to a Double type, you can compare the value with a baseline value. (This is helpful if your task relies on a feature that was introduced in a certain version of SAS.)

Example: Submitting a Program and Displaying the Log

The following C# statements submit a program to the SAS Workspace Server named Local. When the program completes, the log content is displayed in a modal dialog box. The dialog box is an instance of the specialized SASLogViewDialog class in the SAS.Tasks.Toolkit namespace.

```
using SAS.Tasks.Toolkit;
using SAS.Tasks.Toolkit.Controls;

/* following code would appear in a class method */

// helper class to submit a program
SAS.Tasks.Toolkit.SasSubmitter submitter =
    new SAS.Tasks.Toolkit.SasSubmitter("Local");

// hold SAS log output
string log;
```

```
// submit program and wait for completion
submitter.SubmitSasProgramAndWait("proc options; run;",
  out log);

// show the log
SASLogViewDialog dlg =
    new SASLogViewDialog("output",
        "Output of Program", log);
dlg.ShowDialog(this);
```

The log content is displayed in a modal dialog box, similar to Figure 5.1.

Figure 5.1: SAS Log in a Modal Dialog Box

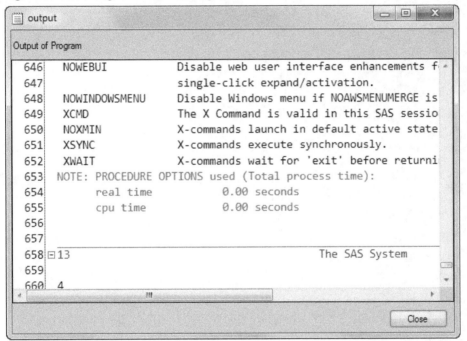

Chapter Summary

It should be clear to you by now—you probably wouldn't want to create a custom task from scratch without the benefit of the Task Toolkit. The Task Toolkit removes much of the tedium from standard task implementations, and it enables you to override behaviors that are necessary for your task. The Task Toolkit contains plenty of helper classes that solve the common challenges associated with interacting with SAS and SAS Enterprise Guide. By taking advantage of built-in tools, you have more time to develop the important features that make your task valuable to your users.

SAS developers who work on SAS Enterprise Guide (including me) use the Task Toolkit for their tasks. That means they are motivated to make the Task Toolkit as solid as possible and to introduce improvements with each new release.

Chapter 6: Your First Custom Task Using Visual Basic

Your First Custom Task Using Visual Basic Express 69
    Creating the Project 70
    Turning the Visual Basic Class into a Custom Task 71
    Build, Deploy, and Test 74
    Adding a User Interface 75
    Saving and Restoring Task Settings... 77
Chapter Summary 81

The best way to learn about custom tasks is to create one. This chapter provides the steps to create a basic custom task using Visual Basic. A novice .NET programmer should be able to complete these steps in less than 30 minutes. Don't worry if you don't understand everything along the way. The concepts are explained in more detail in later chapters.

If you prefer C# as your programming language, stop reading this chapter and move on to Chapter 7, "Your First Custom Task Using C#." Although the steps are very similar for any .NET language, having the steps separated by language should make them easier to follow.

The steps in this chapter assume that you are using Visual Basic 2008 Express, Visual Basic 2010 Express, or Visual Basic 2012 Express. If you are using a Professional Edition of Visual Studio, the screen images might appear slightly different.

Also, the steps in this chapter do not use the Microsoft Visual Studio project templates that are described in Chapter 3, "Creating Custom Task Projects in Microsoft Visual Studio." The templates automate several of the steps in this chapter. So, by not using the templates and following all of the steps, you can learn your way around the software.

Your First Custom Task Using Visual Basic Express

Follow these steps to create a custom task using .NET as your programming language. After each step, the project should build without errors. To build the project, select **Build Solution** from the **Build** menu (or use the keyboard shortcut by holding down the CTRL and SHIFT keys and clicking B).

Creating the Project

Here are the steps to create a basic custom task:

1. Open Visual Basic.
2. Select **File→New Project**. In the New Project dialog box, select **Class Library**. For the project name, enter **BasicStatsVB**. Uncheck the **Create directory for solution** check box. Click **OK**. The project is created in your **My Documents** area.
3. Select **File→Save All**.

   Note: In Chapter 2, "Tools of the Trade," I discussed how different versions of SAS Enterprise Guide support specific versions of the .NET Framework. By default, the later versions of Visual Basic Express target the most recent version of the .NET Framework. This might not be compatible with the version of SAS Enterprise Guide that you're using. For example, Visual Basic 2010 Express targets .NET 4.0, but SAS Enterprise Guide 4.3 supports only .NET 3.5 or earlier.

   The good news is that you can change the version of the .NET Framework that you want to target with your project. That way, you can continue to use the latest version of the .NET Framework to build tasks for earlier versions of SAS Enterprise Guide. The next step describes how to make that change.

4. If your custom task needs to target an earlier version of the .NET Framework (as shown in Table 2.1: SAS Applications and the .NET Framework Versions They Support in Chapter 2), then you need to change the target framework version to build a compatible project. To change the version:

   a. Select **Project→BasicStatsVB Properties**. The project properties appear.
   b. Click the **Application** tab.
   c. Change the value of **Target framework** to the appropriate version of the .NET Framework. For example, for compatibility with SAS Enterprise Guide 4.3, select **NET Framework 3.5**.

      Note: In the earlier versions of Visual Basic Express, you will need to select the **Compile** page, and then **Advanced Compile Options**. The Target Framework setting is in the Advanced Compiler Settings window.

   d. When the confirmation dialog box appears, click **Yes**. The message indicates that this will save, close, and reopen the project to apply the change.
   e. When the project reopens, continue with the remaining steps.

5. In the Solution Explorer view, change the name of the **Class1.vb** file to **BasicStats.vb**. A dialog box asks whether you want to rename all references. Click **Yes**. The class name in the code editor view is renamed automatically to match the new name.
6. In the Solution Explorer view, right-click on the **BasicStatsVB** project, and select **Properties**. The project properties appear. Select **References**. Click **Add**.
7. In the Reference Manager dialog box, click the **Browse** button. Navigate to where SAS Enterprise Guide or the SAS Add-In for Microsoft Office is installed. Select **SAS.Shared.AddIns.dll** and **SAS.Tasks.Toolkit.dll**. You can use the CTRL key to make multiple selections. With both files selected, click **OK**.

8. In the **References** list, highlight these two new references, and view their properties in the Properties view. Change **Copy Local** to **False** (instead of the default value of **True**). This prevents your directory from becoming cluttered with a cascade of dependent .NET assemblies.
9. In the **References** list, click **Add** again. In the Reference Manager dialog box, select **Assemblies→Framework**, and check **System.Windows.Forms**. Click **OK**.
10. Select **Build→Build Solution** to build your project. There should be no errors. If you do encounter errors, go back and review the previous steps carefully. The project must build successfully before you can continue with the remaining steps.

When your project builds successfully, the output is a .NET assembly that contains your class library, which is a DLL file. The DLL file should be in the `bin` subfolder of your project directory (by default, this is something similar to `C:\Documents and Settings\username\My Documents\Visual Studio 2010\Projects\BasicStatsVB\bin\Release`).

Turning the Visual Basic Class into a Custom Task

Now it's time to add code to your task so that it can actually do something.

1. Add the following statements to the top of the BasicStats.vb file, above the **Public Class** statement:
   ```
   Imports SAS.Shared.AddIns
   Imports SAS.Tasks.Toolkit
   ```

2. Add an **Inherits** statement to the end of the BasicStats class declaration:

   ```
   Public Class BasicStats
       Inherits SAS.Tasks.Toolkit.SasTask
   ```

3. Generate a unique identifier for your task. Use the Guidgen tool (or equivalent uuidgen.exe utility). This tool is not directly accessible from the Express editions of Visual Basic. You should be able to find it in `C:\Program Files\Microsoft Visual Studio 9.0\Common7\Tools`. Or, you can use a web-based tool, such as www.guidgen.com.

4. Add the following class attributes at the top of the BasicStats class declaration. The entire class declaration looks similar to this:

   ```
   <ClassId("A9847275-B489-46ef-8FE8-09856A09CD0A"), _
       Version(4.2), _
       InputRequired(InputResourceType.Data)> _
   Public Class BasicStats
       Inherits SAS.Tasks.Toolkit.SasTask
   ```

   Remember, an underscore in Visual Basic signifies a line-continuation character. They are important to include here. In the previous example, you should replace the unique identifier value in the **ClassId** attribute with the value that you generated in the previous step. Do *not* include the brace characters (such as { and }) in the **ClassId** attribute.

5. Save the **BasicStats.vb** file.
6. Because the BasicStats class inherits from the SAS.Tasks.Toolkit.SasTask class, Visual Basic should recognize the class as a designable component. Right-click on the **BasicStats.vb** file, and select **View Designer**. The Design view appears. More importantly, the Properties view includes properties that you can use to describe your task.
7. In the Properties view, enter these values for the properties:
    - For **ProcsUsed**, enter `MEANS`.
    - For **ProductsRequired**, enter `BASE`.
    - For **TaskCategory**, enter `SAS Press Test`.
    - For **TaskDescription**, enter `My first custom task`.
    - For **TaskName**, enter `Basic Statistics`.

    Keep the default values for the remaining properties. When you are finished, the Properties view should look similar to Figure 6.1, if sorted alphabetically.

Figure 6.1: The Properties View of the BasicStats Class

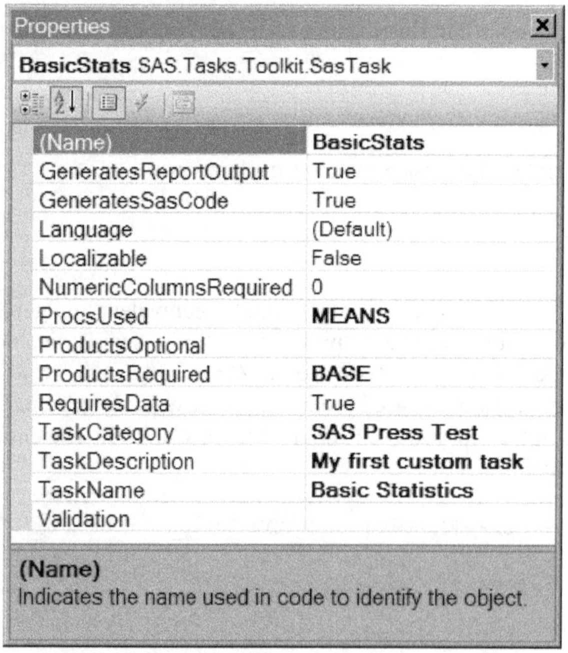

Property	Value
(Name)	BasicStats
GeneratesReportOutput	True
GeneratesSasCode	True
Language	(Default)
Localizable	False
NumericColumnsRequired	0
ProcsUsed	MEANS
ProductsOptional	
ProductsRequired	BASE
RequiresData	True
TaskCategory	SAS Press Test
TaskDescription	My first custom task
TaskName	Basic Statistics
Validation	

(Name)
Indicates the name used in code to identify the object.

8. In the Design view for the BasicStats class, switch to the code editor view by selecting **click here to switch to code view**. In the code editor view, you see the statements that were generated by the new values in the Properties view. A new InitializeComponent routine has been added with statements to set the values that you specified in the previous step.

9. Add a constructor to the BasicStats class. Call the **InitializeComponent** routine. It should look similar to this:

   ```
   Public Sub New()
       InitializeComponent()
   End Sub
   ```

10. Now, override the Show function. A simple way to do this is to begin to enter `Overrides`, and then allow the autocomplete feature of the code editor to lead you to the Show function. Change the body of the **Show** function to return `ShowResult.RunNow`, similar to this:

    ```
    Public Overrides Function Show(ByVal Owner As _
        System.Windows.Forms.IWin32Window) As _
        SAS.Shared.AddIns.ShowResult
        Return ShowResult.RunNow
    End Function
    ```

11. Override the GetSasCode function. Currently, the function returns a simple SAS program that runs the MEANS procedure on the active data. Change the body of the GetSasCode function to look similar to this:

    ```
    Public Overrides Function GetSasCode() As String
      Return String.Format("proc means data={0}.{1}; run;", _
        Consumer.ActiveData.Library, _
        Consumer.ActiveData.Member)
    End Function
    ```

    The entire Visual Basic code for the BasicStats class should look similar to this:

    ```
    Imports SAS.Tasks.Toolkit
    Imports SAS.Shared.AddIns

    <ClassId("A9847275-B489-46ef-8FE8-09856A09CD0A"), _
        Version(4.2), _
        InputRequired(InputResourceType.Data)> _
    Public Class BasicStats
        Inherits SAS.Tasks.Toolkit.SasTask

        Public Sub New()
            InitializeComponent()
        End Sub

        Private Sub InitializeComponent()
            '
            'BasicStats
            '
            Me.ProcsUsed = "MEANS"
            Me.ProductsRequired = "BASE"
            Me.TaskCategory = "SAS Press Test"
            Me.TaskDescription = "My first custom task"
            Me.TaskName = "Basic Statistics"
    ```

```
            End Sub

            Public Overrides Function Show(ByVal Owner As _
                System.Windows.Forms.IWin32Window) As _
                SAS.Shared.AddIns.ShowResult
                Return ShowResult.RunNow
            End Function

            Public Overrides Function GetSasCode() As String
              Return String.Format("proc means data={0}.{1}; run;", _
                  Consumer.ActiveData.Library, _
                  Consumer.ActiveData.Member)
            End Function
        End Class
```

Build, Deploy, and Test

You are now ready to build, deploy, and test your custom task.

1. To build the task, select **Build→Build Solution**. Remember, if you encounter errors, you must correct them before you can deploy and test the task.
2. To deploy the task for use in either SAS Enterprise Guide or the SAS Add-In for Microsoft Office, copy the DLL file from the project output directory (`\bin\Release\BasicStatsVB.dll`) to the local `Custom` task folder (`%appdata%\SAS\SharedSettings\4.3\Custom`).

   For more information about how to deploy a custom task, see the section, "Deploying Custom Tasks," in Chapter 1, "Why Custom Tasks."

To test the task in SAS Enterprise Guide:

1. Open SAS Enterprise Guide.
2. Select **Tools→Add-In→Basic Statistics**.
3. In the Open dialog box, select a data source from a SAS library, and click **OK**. SAS Enterprise Guide adds the **Basic Statistics** task to your project, and then runs the MEANS procedure with the selected data source.

To test the task in the SAS Add-In for Microsoft Office (using Microsoft Excel):

1. Open Microsoft Excel.
2. Click the **SAS** tab. Select the **Tasks** menu. The **Tasks** menu options include tasks organized into categories.
3. Select **Basic Statistics**. (In some versions of the SAS Add-In, this task will appear in the **SAS Press Test** category.)
4. In the Open dialog box, select a data source from a SAS library or from Microsoft Excel, and click **OK**. The SAS Add-In for Microsoft Office adds the **Basic Statistics** task to your project, and then runs the MEANS procedure with the selected data source. The results are saved in your workbook.

Adding a User Interface

At this point, you have a completely functional custom task. It is available in the menu for tasks in the different applications. You can run it and see it generate results. But, it is missing some fundamental features. For example, the task has no user interface. Without a user interface, the task is doomed to perform the same dull process each time it is run. There is no opportunity for user interaction.

In this section, you will add a basic user interface that enables you to control some of the task's behavior. The user interface includes a single Windows Form control with some statistics that are included in the results. It also includes the option to specify a title for the output.

Here are the steps for adding a Windows Form to your task:

1. In the Solution Explorer view, right-click on **BasicStatsVB**, and select **Add→Windows Form**. The Add New Item dialog box appears.
2. Enter **BasicStatsForm.vb** as the name for the new Windows Form file, and click **Add**. The blank BasicStatsForm appears in the Design view, as shown in Figure 6.2.

Figure 6.2: The Blank BasicStatsForm for the Task

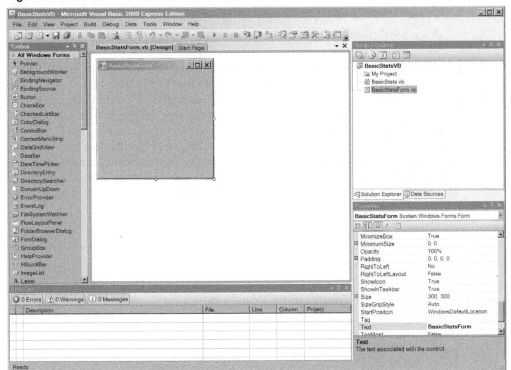

3. From the Toolbox view, under **Common Controls**, drag and drop two **Button** controls onto the BasicStatsForm canvas. Position them next to each other in the bottom right corner. In the Properties view for each button, enter these values for the properties:
    - For the button on the right, change **Text** to `Cancel`, **Name** to `btnCancel`, and **DialogResult** to `Cancel`.
    - For the button on the left, change **Text** to `OK`, **Name** to `btnOK`, and **DialogResult** to `OK`.
    - For both buttons, change **Anchor** to **Bottom, Right**.

4. Click in the title bar of the BasicStatsForm in the Design view to select the form's properties. In the Properties view, enter these values for the properties:
    - Change **AcceptButton** to `btnOK`.
    - Change **CancelButton** to `btnCancel`.
    - Change **MaximizeBox** to `False`.
    - Change **MinimizeBox** to `False`.
    - Change **MinimumSize** to `300,200`.
    - Change **ShowInTaskbar** to `False`.
    - Change **StartPosition** to `CenterParent`.

5. From the Toolbox view, under **Common Controls**, drag and drop a **Label** control onto the BasicStatsForm canvas. In the Properties view, set **Text** to `Title:`.
6. From the Toolbox view, drag and drop a **TextBox** control onto the BasicStatsForm canvas. Position it below the **Label** control. In the Properties view, set **Name** to `txtTitle` and **Anchor** to **Top, Left, Right**.
7. From the Toolbox view, drag and drop a **CheckBox** control onto the BasicStatsForm canvas. Position it below the **TextBox**. In the Properties view, set **CheckBox** name to **chkIncludeModeMedian**. Set **Text** to **Include mode and median**. At this point, the BasicStatsForm in the Design view should look similar to Figure 6.3.

Figure 6.3: BasicStatsForm with Controls

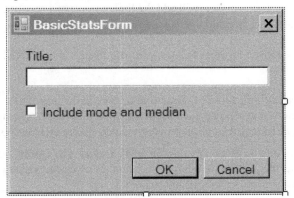

8. In the BasicStats.vb file, add the following statement to the top of the file:

   Imports System.Windows.Forms
9. In the Show function in the BasicStats.vb file, change the body of the function to set the title of the BasicStatsForm and to check for the `DialogResult` value:

```
Public Overrides Function Show(ByVal Owner As _
    System.Windows.Forms.IWin32Window) As _
    SAS.Shared.AddIns.ShowResult

    Dim dlg As BasicStatsForm = New BasicStatsForm()
    ' set title for form
    dlg.Text = String.Format("Basic Stats for {0}.{1}", _
      Consumer.ActiveData.Library, _
      Consumer.ActiveData.Member)

    ' show form
    If (DialogResult.OK = dlg.ShowDialog(Owner)) Then
    ' save the settings from the form
      Return ShowResult.RunNow
    Else : Return ShowResult.Canceled
    End If
End Function
```

The user interface is now completed. You can build your project and deploy the .NET assembly for testing. When you run the task, the BasicStatsForm is displayed, enabling you to change the title and to include the mode and median statistics in your output. However, you have not added any code to save these preferences or to generate the appropriate SAS program to apply them to. This remaining gap brings us to the last set of steps to make this an *even better* completely functional custom task.

Saving and Restoring Task Settings

The task is showing real promise. You can use it to run a basic analysis, and it provides users with some control over the results. (Well, at least it provides them the illusion of control.) Because the task does not yet recognize the user-supplied settings from BasicStatsForm, the user will probably be frustrated with the lack of real control. In this section, you will add the logic to save and restore the task settings and to generate an appropriate SAS program that is based on these settings.

There are several ways to save and restore results. In this example, you will create a special .NET class that serves several purposes:

- Contains properties that represent the structure of information to store in the task. These properties correspond to the options that are displayed in the user interface. In this example, there is a text property named **Title**, and a Boolean property named **IncludeModeMedian**.
- Provides methods to save and restore the property values in XML format. The custom task APIs are designed to use XML as the persistence (storage) medium to save settings between task runs. In this example, these methods are **ToXml** and **FromXml**.
- Provides a method to generate an appropriate SAS program that is based on the task settings. In this example, this method is named **ToSasProgram**.

To add the Settings class to your project:

1. In the Solution Explorer view, right-click on **BasicStatsVB**, and select **Add→New Item**.
2. In the Add New Item dialog box, select **Class**. Name the class **BasicStatsSettings.vb**. Click **Add**. The new class is displayed in the code editor view.
3. Replace the existing contents of the BasicStatsSettings.vb file with the following:

```
Imports System.Xml
Imports System.Text

Public Class BasicStatsSettings
  Public IncludeModeMedian As Boolean
  Public Title As String

  Public Function ToXml() As String
    Dim doc As New XmlDocument
    Dim el As XmlElement
    el = doc.CreateElement("BasicStats")
    el.SetAttribute("IncludeModeMedian", _
      XmlConvert.ToString(IncludeModeMedian))
    el.SetAttribute("Title", Title)
    doc.AppendChild(el)
    Return doc.OuterXml
  End Function

  Public Sub FromXml(ByVal savedXml As String)
    Dim doc As New XmlDocument
    doc.LoadXml(savedXml)
    Dim el As XmlElement
    el = doc("BasicStats")
    IncludeModeMedian = _
```

```
      XmlConvert.ToBoolean( _
        el.Attributes("IncludeModeMedian").Value)
    Title = el.Attributes("Title").Value
  End Sub

  Public Function ToSasProgram(ByVal data As String) As String
    Dim sb As New StringBuilder()
    sb.AppendFormat("title '{0}';", Title)
    sb.AppendLine()
    sb.AppendFormat("proc means data={0}", data)
    If (IncludeModeMedian) Then
      sb.AppendLine()
      sb.AppendLine("    n sum mean std mode median;")
    Else : sb.AppendLine(";")
    End If
    sb.AppendLine("run;")
    Return sb.ToString()
  End Function
End Class
```

4. To teach the task's form to initialize itself with these saved settings, change the constructor of the BasicStatsForm class in the BasicStatsForm.vb file to accept a value of the BasicStatsSettings type. At the same time, transfer the values from the task's form back to the Settings class when the form is closing. Change the body of the BasicStatsForm class to the following:

```
Public Settings As BasicStatsSettings

Public Sub New(ByVal settings As BasicStatsSettings)
  InitializeComponent()
  Me.Settings = settings
  txtTitle.Text = settings.Title
  chkIncludeMode.Checked = settings.IncludeModeMedian
End Sub

Protected Overrides Sub OnClosed( _
  ByVal e As System.EventArgs)
  If (DialogResult = Windows.Forms.DialogResult.OK) Then
    Settings.IncludeModeMedian = chkIncludeMode.Checked
    Settings.Title = txtTitle.Text
  End If
  MyBase.OnClosed(e)
End Sub
```

5. In the BasicStats class in the BasicStats.vb file, add a member variable to hold the value for the Settings object and to initialize it appropriately. Add the following to the body of the BasicStats class:

```
Private Settings As New BasicStatsSettings
```

6. Replace the code that generates the SAS program with a call to the ToSasProgram method in the BasicStatsSettings class. Change the body of the GetSasCode method to the following:

```
Public Overrides Function GetSasCode() As String
  Return settings.ToSasProgram(String.Format("{0}.{1}", _
    Consumer.ActiveData.Library, _
    Consumer.ActiveData.Member))
End Function
```

7. To save and restore the XML state of the task, add the GetXmlState and RestoreStateFromXml methods. Add the following to the body of the BasicStats class:

```
Public Overrides Sub RestoreStateFromXml(ByVal xmlState _
  As String)
    settings.FromXml(xmlState)
End Sub

Public Overrides Function GetXmlState() As String
  Return settings.ToXml()
End Function
```

8. Change the BasicStatsForm constructor in the Show function to pass in the instance of the BasicStatsSettings class to initialize the form. The body of the Show function should look similar to this:

```
Public Overrides Function Show(ByVal Owner As _
  System.Windows.Forms.IWin32Window) As _
  SAS.Shared.AddIns.ShowResult

  Dim dlg As BasicStatsForm = New BasicStatsForm(settings)
  ' set title for form
  dlg.Text = String.Format("Basic Stats for {0}.{1}", _
    Consumer.ActiveData.Library, _
    Consumer.ActiveData.Member)

  ' show form
  If (DialogResult.OK = dlg.ShowDialog(Owner)) Then
    ' save the settings from the form
    settings = dlg.Settings
    Return ShowResult.RunNow
  Else : Return ShowResult.Canceled
  End If
End Function
```

The coding of the task is now completed. You can build, deploy, and test the task as you did in earlier steps. You should verify that when you select the check box to include the mode and median and you enter a title, the correct SAS program is generated, and the settings are remembered when you rerun it the next time. Figure 6.4 shows an example of the task in action.

Figure 6.4: The Completed Basic Stats Task

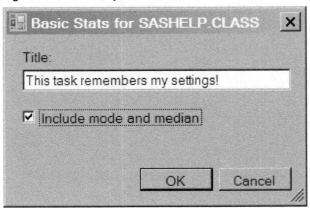

Chapter Summary

If you followed the steps in this chapter, you now have your first working task implemented in Visual Basic. Even if you don't understand all of the technical reasons for the steps that you performed, that's okay. The nitty gritty details are covered in later chapters. At this point, you have learned something about how to use the development tools and you understand the mechanics of building and deploying tasks for use.

82

Chapter 7: Your First Custom Task Using C#

Your First Custom Task Using Visual C# Express 83
    Creating the Project 84
    Turning the C# Class into a Custom Task 85
    Build, Deploy, and Test 89
    Adding a User Interface 89
    Saving and Restoring Task Settings 92
Chapter Summary 96

The best way to learn about custom tasks is to create one. This chapter provides the steps to create a basic custom task using C#. A novice .NET programmer should be able to complete these steps in less than 30 minutes. Don't worry if you don't understand everything along the way. The concepts are explained in more detail in later chapters.

If you prefer Visual Basic as your programming language, stop reading this chapter and go back to Chapter 6, "Your First Custom Task Using Visual Basic." Although the steps are very similar for any .NET language, having the steps separated by language should make them easier to follow.

The steps in this chapter assume that you are using Visual C# 2008, Visual C# 2010 Express, or Visual C# 2012 Express. If you are using a Professional Edition of Visual Studio, the screen images might appear slightly different.

Also, the steps in this chapter do not use the Microsoft Visual Studio project templates that are described in Chapter 3, "Creating Custom Task Projects in Microsoft Visual Studio." The templates automate several of the steps in this chapter. So, by not using the templates and following all of the steps, you can learn your way around the software.

Your First Custom Task Using Visual C# Express

Follow these steps to create a custom task using C# as your programming language. After each step, the project should build without errors. To build the project, select **Build Solution** from the **Build** menu (or use the keyboard shortcut by holding down the CTRL and SHIFT keys and clicking B).

Creating the Project

Here are the steps to create a basic custom task:

1. Open Visual C#.
2. Select **File→New Project**. In the New Project dialog box, select **Class Library**. For the project name, enter **BasicStatsCSharp**. Uncheck the **Create directory for solution** check box. Click **OK**. The project is created in your **My Documents** area.
3. Select **File→Save All**.

   Note: In Chapter 2, "Tools of the Trade," I discussed how different versions of SAS Enterprise Guide support specific versions of the .NET Framework. By default, the later versions of Visual C# Express target the most recent version of the .NET Framework. This might not be compatible with the version of SAS Enterprise Guide that you're using. For example, Visual C# 2010 Express targets .NET 4.0, but SAS Enterprise Guide 4.3 supports only .NET 3.5 or earlier.

   The good news is that you can change the version of the .NET Framework that you want to target with your project. That way, you can continue to use the latest version of the .NET Framework to build tasks for earlier versions of SAS Enterprise Guide. The next step describes how to make that change.

4. If your custom task needs to target *an earlier version of the .NET Framework* (as shown in Table 2.1: SAS Applications and the .NET Framework Versions They Support in Chapter 2), then you need to change the target framework version to build a compatible project. To change the version:

   a. Select **Project→BasicStatsCSharp Properties**. The project properties appear. Confirm the name and directory for your project.
   b. Click the **Application** tab.
   c. Change the value of **Target framework** to the appropriate version of the .NET Framework. For example, for compatibility with SAS Enterprise Guide 4.3, select **NET Framework 3.5**.
   d. When the confirmation dialog box appears, click **Yes**. The message indicates that this will save, close, and reopen the project to apply the change.
   e. When the project reopens, continue with the remaining steps.

5. In the Solution Explorer view, change the name of the **Class1.cs** file to **BasicStats.cs**. A dialog box asks whether you want to rename all references. Click **Yes**. The class name in the code editor view is renamed automatically to match the new name.
6. In the Solution Explorer view, right-click on **References**. Select **Add Reference**.
7. In the Add Reference dialog box, click the **Browse** tab. Navigate to where SAS Enterprise Guide or the SAS Add-In for Microsoft Office is installed. Select **SAS.Shared.AddIns.dll** and **SAS.Tasks.Toolkit.dll**. You can use the CTRL key to make multiple selections. With both files selected, click **OK**.
8. In the **References** list, highlight these two new references, and view their properties in the Properties view. Change **Copy Local** to **False** (instead of the default value of **True**). This

prevents your directory from becoming cluttered with a cascade of dependent .NET assemblies.
9. In the Solution Explorer view, right-click on **References**. Select **Add Reference**. In the Add Reference dialog box, click the **.NET** tab, and select **System.Windows.Forms**. Click **OK**.
10. Select **Build**→**Build Solution** to build your project. (Note: in some versions of Visual C# Express, you might need to select **Tools**→**Settings**→**Expert Settings** before you can see the **Build** menu.) There should be no errors. If you do encounter errors, go back and review the previous steps carefully. The project must build successfully before you can continue with the remaining steps.

When your project builds successfully, the output is a .NET assembly that contains your class library, which is a DLL file. The DLL file should be in the `bin` subfolder of your project directory (by default, this is something similar to `C:\Documents and Settings\username\My Documents\Visual Studio 2010\Projects\BasicStatsCSharp\bin\Release`).

Turning the C# Class into a Custom Task

Now it's time to add code to your task so that it can actually do something.

1. Add the following statements to the very top of the **BasicStats.cs** file:
    ```
    using SAS.Shared.AddIns;
    using SAS.Tasks.Toolkit;
    ```
2. Add a statement to the BasicStats class declaration so that it inherits from the SAS.Tasks.Toolkit.SasTask class, similar to this:
    ```
    public class BasicStats : SAS.Tasks.Toolkit.SasTask
    ```
3. Generate a unique identifier for your task. Use the Guidgen tool (or equivalent uuidgen.exe utility). This tool is not directly accessible from the Express editions of Visual C#. You should be able to find it in `C:\Program Files\Microsoft Visual Studio 9.0\Common7\Tools`. Or, you can use a web-based tool, such as www.guidgen.com.
4. Add the following class attributes at the top of the BasicStats class declaration. The entire class declaration looks similar to this:
    ```
    [ClassId("0f6d3598-23ce-43b6-8178-5703cbe094f5")]
    [Version(4.2)]
    [InputRequired(InputResourceType.Data)]
    public class BasicStats : SAS.Tasks.Toolkit.SasTask
    {
    }
    ```

    In the previous example, you should replace the unique identifier value in the **ClassId** attribute with the value that you generated in the previous step. Do *not* include the brace characters (such as { and }) in the **ClassId** attribute.
5. Save the **BasicStats.cs** file.

6. Add two methods to the body of the BasicStats class. One method is **InitializeComponent**, which holds code that is generated in the next step. The other method is a default constructor, which includes a call to **InitializeComponent**. When you are finished, the code for the BasicStats class should look similar to this:

```
[ClassId("0f6d3598-23ce-43b6-8178-5703cbe094f5")]
[Version(4.2)]
[InputRequired(InputResourceType.Data)]
public class BasicStats : SAS.Tasks.Toolkit.SasTask
{
    public BasicStats()
    {
        InitializeComponent();
    }

    private void InitializeComponent()
    {
    }
}
```

7. Because the BasicStats class inherits from the SAS.Tasks.Toolkit.SasTask class, Visual C# should recognize the class as a designable component. Right-click on the **BasicStats.cs** file, and select **View Designer**. The Design view appears. More importantly, the Properties view includes properties that you can use to describe your task.

8. In the Properties view, enter these values for the properties:

   o For **ProcsUsed**, enter `MEANS`.

   o For **ProductsRequired**, enter `BASE`.

   o For **TaskCategory**, enter `SAS Press Test`.

   o For **TaskDescription**, enter `My first custom task`.

   o For **TaskName**, enter `Basic Statistics`.

   Keep the default values for the remaining properties. When you are finished, the Properties view should look similar to Figure 7.1, if sorted alphabetically.

Figure 7.1: The Properties View of the BasicStats class

9. Now, override the Show function. A simple way to do this is to begin to enter **Override**, and then allow the autocomplete feature of the code editor to lead you to the Show function. Change the body of the Show function to return ShowResult.RunNow, similar to this:

   ```
   public override ShowResult Show(
       System.Windows.Forms.IWin32Window Owner)
   {
       return ShowResult.RunNow;
   }
   ```

10. Override the GetSasCode function. Currently, the function returns a simple SAS program that runs the MEANS procedure on the active data. Change the body of the GetSasCode function to look similar to this:

    ```
    public override string GetSasCode()
    {
        return string.Format("proc means data={0}.{1}; run;",
            Consumer.ActiveData.Library,
            Consumer.ActiveData.Member);
    }
    ```

The entire Visual C# code for the BasicStats class should look similar to this:

```csharp
using System;
using System.Collections.Generic;
using System.Linq;
using System.Text;
using SAS.Shared.AddIns;
using SAS.Tasks.Toolkit;

namespace BasicStatsCSharp
{
  [ClassId("0f6d3598-23ce-43b6-8178-5703cbe094f5")]
  [Version(4.2)]
  [InputRequired(InputResourceType.Data)]
  public class BasicStats : SAS.Tasks.Toolkit.SasTask
  {
    public BasicStats()
    {
      InitializeComponent();
    }

    private void InitializeComponent()
    {
      //
      // BasicStats
      //
      this.ProcsUsed = "MEANS";
      this.ProductsRequired = "BASE";
      this.TaskCategory = "SAS Press Test";
      this.TaskDescription = "My first custom task";
      this.TaskName = "Basic Statistics";
    }

    public override ShowResult Show(
      System.Windows.Forms.IWin32Window Owner)
    {
      return ShowResult.RunNow;
    }

    public override string GetSasCode()
    {
      return string.Format("proc means data={0}.{1}; run;",
        Consumer.ActiveData.Library,
        Consumer.ActiveData.Member);
    }
  }
}
```

Build, Deploy, and Test

You are now ready to build, deploy, and test your custom task.

1. To build the task, select **Build**→**Build Solution**. Remember, if you encounter errors, you must correct them before you can deploy and test the task.
2. To deploy the task for use in either SAS Enterprise Guide or the SAS Add-In for Microsoft Office, copy the DLL from the project output directory (`\bin\Release\BasicStatsCSharp.dll`) to the local `Custom` task folder (`%appdata%\SAS\SharedSettings\4.3\Custom` or `%appdata%\SAS\SharedSettings\5.1\Custom`, depending on which version of the SAS applications you have installed).

   For more information about how to deploy a custom task, see the section, "Deploying Custom Tasks," in Chapter 1, "Why Custom Tasks."

To test the task in SAS Enterprise Guide:

1. Open SAS Enterprise Guide.
2. Select **Tools**→**Add-In**→**Basic Statistics**.
3. In the Open dialog box, select a data source from a SAS library. and click **OK**. SAS Enterprise Guide adds the **Basic Statistics** task to your project, and then runs the MEANS procedure with the selected data source.

To test the task in the SAS Add-In for Microsoft Office (using Microsoft Excel):

1. Open Microsoft Excel.
2. Click the **SAS** tab. Select the **Tasks** menu. The **Tasks** menu options include tasks organized into categories.
3. Select **Basic Statistics**. (Note: in some versions of the SAS Add-In, the task will appear in the **SAS Press Test** category.
4. In the Open dialog box, select a data source from a SAS library or from Microsoft Excel, and click **OK**. The SAS Add-In for Microsoft Office adds the **Basic Statistics** task to your project, and then runs the MEANS procedure with the selected data source. The results are saved in your workbook.

Adding a User Interface

At this point, you have a completely functional custom task. It is available in the menu for tasks in the different applications. You can run it and see it generate results. But, it is missing some fundamental features. For example, the task has no user interface. Without a user interface, the task is doomed to perform the same dull process each time it is run. There is no opportunity for user interaction.

In this section, you will add a basic user interface that enables you to control some of the task's behavior. The user interface includes a single Windows Form control with some statistics that are included in the results. It also includes the option to specify a title for the output.

Here are the steps for adding a Windows Form to your task:

1. In the Solution Explorer view, right-click on **BasicStatsCSharp**, and select **Add→Windows Form**. The Add New Item dialog box appears.
2. Enter **BasicStatsForm.cs** as the name for the new Windows Form file, and click **Add**. The blank BasicStatsForm appears in the Design view, as shown in Figure 7.2.

Figure 7.2: The Blank BasicStatsForm for the Task

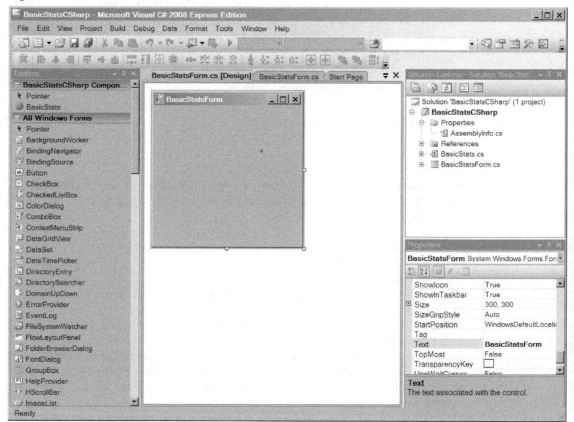

3. From the Toolbox view, under **Common Controls**, drag and drop two **Button** controls onto the BasicStatsForm canvas. Position them next to each other in the bottom right corner. In the Properties view for each button, enter these values for the properties:
    - For the button on the right, change **Text** to `Cancel`, **Name** to `btnCancel`, and **DialogResult** to `Cancel`.
    - For the button on the left, change **Text** to `OK`, **Name** to `btnOK`, and **DialogResult** to `OK`.
    - For both buttons, change **Anchor** to **Bottom, Right**.

4. Click in the title bar of the BasicStatsForm in the Design view to select the form's properties. In the Properties view, enter these values for the properties:
   - Change **AcceptButton** to `btnOK`.
   - Change **CancelButton** to `btnCancel`.
   - Change **MaximizeBox** to `False`.
   - Change **MinimizeBox** to `False`.
   - Change **MinimumSize** to `300,200`.
   - Change **ShowInTaskbar** to `False`.
   - Change **StartPosition** to `CenterParent`.
5. From the Toolbox view, under **Common Controls**, drag and drop a **Label** control onto the BasicStatsForm canvas. In the Properties view, set **Text** to `Title:`.
6. From the Toolbox view, drag and drop a **TextBox** control onto the BasicStatsForm canvas. Position it below the **Label** control. In the Properties view, set **Name** to `txtTitle` and **Anchor** to **Top, Left, Right**.
7. From the Toolbox view, drag and drop a **CheckBox** control onto the BasicStatsForm canvas. Position it below the **TextBox**. In the Properties view, set **CheckBox** name to `chkIncludeModeMedian`. Set **Text** to `Include mode and median`. At this point, the BasicStatsForm in the Design view should look similar to Figure 7.3.

Figure 7.3: BasicStatsForm with Controls

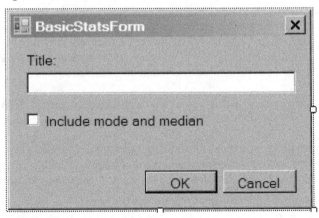

8. In the BasicStats.cs file, add the following statement to the top of the file (make sure you are in code editor view):

    ```
    using System.Windows.Forms;
    ```

9. In the Show function in the BasicStats.cs file, change the body of the function to set the title of the BasicStatsForm and to check for the `DialogResult` value:

```
public override ShowResult Show(
    System.Windows.Forms.IWin32Window Owner)
{
    BasicStatsForm dlg = new BasicStatsForm();
    // set the title of the form
    dlg.Text = string.Format("Basic Stats for {0}.{1}",
        Consumer.ActiveData.Library,
        Consumer.ActiveData.Member);
    // show the form
    if (DialogResult.OK == dlg.ShowDialog(Owner))
    {
        // if okay, then run task
        return ShowResult.RunNow;
    }
    else
        // don't run task, cancel without saving
        return ShowResult.Canceled;
}
```

The user interface is now completed. You can build your project and deploy the .NET assembly for testing. When you run the task, the BasicStatsForm is displayed, enabling you to change the title and to include the mode and median statistics in your output. However, you have not added any code to save these preferences or to generate the appropriate SAS program to apply them to. This remaining gap brings us to the last set of steps to make this an *even better* completely functional custom task.

Saving and Restoring Task Settings

The task is showing real promise. You can use it to run a basic analysis, and it provides users with some control over the results. (Well, at least it provides them the illusion of control.) Because the task does not yet recognize the user-supplied settings from BasicStatsForm, the user will probably be frustrated with the lack of real control. In this section, you will add the logic to save and restore the task settings and to generate an appropriate SAS program that is based on these settings.

There are several ways to save and restore results. In this example, you will create a special .NET class that serves several purposes:

- Contains properties that represent the structure of information to store in the task. These properties correspond to the options that are displayed in the user interface. In this example, there is a text property named **Title**, and a Boolean property named **IncludeModeMedian**.

- Provides methods to save and restore the property values in XML format. The custom task APIs are designed to use XML as the persistence (storage) medium to save settings between task runs. In this example, these methods are **ToXml** and **FromXml**.

- Provides a method to generate an appropriate SAS program that is based on the task settings. In this example, this method is named **ToSasProgram**.

To add the Settings class to your project:

1. In the Solution Explorer view, right-click on **BasicStatsCSharp**, and select **Add→New Item**.
2. In the Add New Item dialog box, select **Class**. Name the class **BasicStatsSettings.cs**. Click **Add**. The new class is displayed in the code editor view.
3. Replace the existing contents of the BasicStatsSettings.cs file with the following:

```csharp
using System;
using System.Collections.Generic;
using System.Linq;
using System.Text;
using System.Xml;

namespace BasicStatsCSharp
{
  public class BasicStatsSettings
  {
    #region Properties

    public bool IncludeModeMedian { get; set; }
    public string Title { get; set; }

    #endregion

    #region serialization to/from XML
    public string ToXml()
    {
      XmlDocument doc = new XmlDocument();
      XmlElement el = doc.CreateElement("BasicStats");
      el.SetAttribute("IncludeModeMedian",
         XmlConvert.ToString(IncludeModeMedian));
      el.SetAttribute("Title", Title);
      doc.AppendChild(el);
      return doc.OuterXml;
    }

    public void FromXml(string savedXml)
    {
      XmlDocument doc = new XmlDocument();
      doc.LoadXml(savedXml);
      XmlElement el = doc["BasicStats"];
      IncludeModeMedian =
         XmlConvert.ToBoolean(
           el.Attributes["IncludeModeMedian"].Value);
      Title = el.Attributes["Title"].Value;
    }
    #endregion

    #region SAS code generation
    public string ToSasProgram(string data)
```

```
        {
          StringBuilder sb = new StringBuilder();
          sb.AppendFormat("title '{0}';", Title);
          sb.AppendLine();
          sb.AppendFormat("proc means data={0}", data);
          if (IncludeModeMedian)
          {
            sb.AppendLine();
            sb.AppendLine("   n sum mean std mode median;");
          }
          else
            sb.AppendLine(";");
          sb.AppendLine("run;");
          return sb.ToString();
        }
        #endregion
      }
    }
```

4. To teach the task's form to initialize itself with these saved settings, change the constructor of the BasicStatsForm class in the BasicStatsForm.cs file to accept a value of the BasicStatsSettings type. At the same time, transfer the values from the task's form back to the Settings class when the form is closing. Change the body of the BasicStatsForm class to the following:

```
using System;
using System.Collections.Generic;
using System.ComponentModel;
using System.Data;
using System.Drawing;
using System.Linq;
using System.Text;
using System.Windows.Forms;

namespace BasicStatsCSharp
{
  public partial class BasicStatsForm : Form
  {
    internal BasicStatsSettings Settings { get; set; }

    public BasicStatsForm(BasicStatsSettings settings)
    {
      InitializeComponent();
      this.Settings = settings;
      txtTitle.Text = Settings.Title;
      chkIncludeModeMedian.Checked =
        Settings.IncludeModeMedian;
    }

    protected override void OnClosed(EventArgs e)
    {
```

```
          if (DialogResult == DialogResult.OK)
          {
            Settings.IncludeModeMedian =
               chkIncludeModeMedian.Checked;
            Settings.Title = txtTitle.Text;
          }
          base.OnClosed(e);
       }
     }
   }
```

5. In the BasicStats class in the BasicStats.cs file, add a member variable to hold the value for the Settings object and to initialize it appropriately. Add the following to the body of the BasicStats class:

   ```
   BasicStatsSettings settings = new BasicStatsSettings();
   ```

6. Replace the code that generates the SAS program with a call to the ToSasProgram method in the BasicStatsSettings class. Change the body of the GetSasCode method to the following:

   ```
   public override string GetSasCode()
   {
     return settings.ToSasProgram(string.Format("{0}.{1}",
       Consumer.ActiveData.Library,
       Consumer.ActiveData.Member));
   }
   ```

7. To save and restore the XML state of the task, add the GetXmlState and RestoreStateFromXml methods. Add the following to the body of the BasicStats class:

   ```
   public override void RestoreStateFromXml(string xmlState)
   {
     settings.FromXml(xmlState);
   }

   public override string GetXmlState()
   {
     return settings.ToXml();
   }
   ```

8. Change the BasicStatsForm constructor in the Show function to pass in the instance of the BasicStatsSettings class to initialize the form. The body of the Show function should look similar to this:

   ```
   public override ShowResult Show(
     System.Windows.Forms.IWin32Window Owner)
   {
     BasicStatsForm dlg = new BasicStatsForm(settings);
     // set the title of the form
     dlg.Text = string.Format("Basic Stats for {0}.{1}",
       Consumer.ActiveData.Library,
       Consumer.ActiveData.Member);
   ```

```
      // show the form
      if (DialogResult.OK == dlg.ShowDialog(Owner))
      {
        // if okay, then save settings and run task
        settings = dlg.Settings;
        return ShowResult.RunNow;
      }
      else
        // don't run task, cancel without saving
        return ShowResult.Canceled;
}
```

The coding of the task is now completed. You can build, deploy, and test the task as you did in earlier steps. You should verify that when you select the check box to include the mode and median and you enter a title, the correct SAS program is generated, and the settings are remembered when you rerun it the next time. Figure 7.4 shows an example of the task in action.

Figure 7.4: The Completed Basic Stats Task

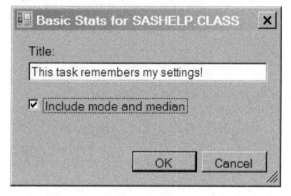

Chapter Summary

If you followed the steps in this chapter, you now have your first working task implemented in Visual C#. Even if you don't understand all of the technical reasons for the steps that you performed, that's okay. The nitty gritty details are covered in later chapters. At this point, you have learned something about how to use the development tools and you understand the mechanics of building and deploying tasks for use.

Chapter 8: Debugging Techniques: Yes, You Will Need Them

Best Practices for Making Your Software Debuggable 98
    Take Advantage of Object-Oriented Design 98
    Consider Unit Testing 99
    Catch and Handle Exceptions 100
    Use Logging to Record Events and Progress 106
Debug with Microsoft Visual Studio 111
    Debugging Basics: Some Definitions 111
    Prepare to Debug a Custom Task 112
    How to Attach a Debugger to a Custom Task 112
    Example: Debugging a Custom Task 113
Chapter Summary 116

Perhaps you are one of those rare developers who can craft a body of code, build it, and deploy it, without encountering a single error or unexpected condition. But, just in case that's not you, this chapter describes the basics of troubleshooting problems in your custom tasks.

You might have heard the life-enhancing advice, "Dance as if nobody is watching." Among software developers, there's a mantra that's almost the opposite, "Code as if bugs are always lurking." This mantra means that regardless of how careful you are, your software will run into bugs, whether they are conditions that you didn't expect or logic errors in your code. The best that you can do is to write your code so that conditions are easy to diagnose and errors are minimized.

Best Practices for Making Your Software Debuggable

Although it can be very difficult to write bug-free code, there are a few proven practices that can help you locate and squash bugs quickly and with confidence. Here are the leading practices that you can apply to custom tasks:

- Take advantage of object-oriented design.
- Consider unit testing.
- Catch and handle exceptions.
- Use logging to record events and progress.

In the following sections, each of these practices is addressed. This is not an exhaustive list! In the discipline of software development, there are dozens of methodologies and tools that programmers have found to be effective. The approach that works best for you depends on the complexity of your task, the dynamics of the team that you work with, and your general tolerance for using methodologies and tools.

Take Advantage of Object-Oriented Design

In an object-oriented design model, software instructions are organized into logical classes. Each class has properties and methods that help the object fulfill its purpose. And, class definitions can inherit from each other. This enables you to organize similar tasks so that shared routines are in a central location, not duplicated across different projects.

When you develop custom tasks using Microsoft .NET, you're using object-oriented design. You don't really have a choice because the provided APIs and Task Toolkit are structured as a collection of objects. For example, the SAS.Tasks.Toolkit.SasTask class *encapsulates* the properties and behaviors that are common to the majority of custom tasks. Because your task class *inherits* from the SasTask class, it has access to these properties and behaviors without requiring you to implement each one yourself. However, if you need to, you can *override* the properties and behaviors in your task.

Different tasks can have different behaviors even though SAS Enterprise Guide (the application that drives them) doesn't know about these different behaviors in advance. A calling function from SAS Enterprise Guide treats different task classes in exactly the same way even though they have different behaviors. That concept is called *polymorphism*.

Inheritance, encapsulation, and *polymorphism* are all fundamental aspects of object-oriented design. That's about as deep as I will get into the theory in this book. Let's look at some examples to see how the theory works in practice.

Figure 8.1 shows a class diagram of the task windows that are used in the Macro Variable Viewer task and the Options Value Viewer task. These classes have much in common with each other. It makes sense to consolidate the common behaviors into a single class, and then to create specialized classes in each task for only the behaviors that must be different.

Figure 8.1: Example of Class Inheritance

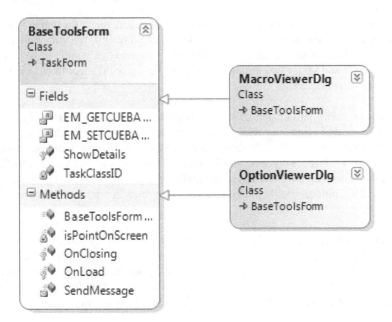

In this example, the **BaseToolsForm** class takes care of a few behaviors that each derived class needs. For example, the **BaseToolsForm** class saves task settings such as window location and size when the task is closed. It restores them the next time the task is opened. These behaviors are captured in the shared **Fields** and **Methods** that you see in the **BaseToolsForm** box on the left.

Notice how the **BaseToolsForm** class actually inherits from *another* class: SAS.Tasks.Toolkit.Controls.TaskForm. As part of the Task Toolkit, the TaskForm class implements several behaviors that are common to almost all custom tasks.

Consider Unit Testing

One of the challenges of creating custom tasks for SAS Enterprise Guide is that there is really only one way to test them—run them in SAS Enterprise Guide.

This can be a cumbersome exercise. Depending on the nature of the task, a thorough test can involve accessing a SAS environment, connecting to databases, and processing SAS programs. If you are part of a large SAS ecosystem with many users and an IT staff, the SAS administrators might not appreciate the iterative nature of your code, compile, and run cycles. This is a lot of spinning up of connections and processing only to test the latest tweak to your task.

Unit testing is the practice of devising your code so that it can be tested independent of the larger infrastructure that is required to run it in production. You write additional code as unit tests. The unit tests should be designed to exercise your methods and classes and make sure they perform as designed.

You use a *unit testing framework* to run these tests, capture the results, and verify the actual results against the expected results. When the actual results match the expected results, the unit test has passed. When they don't match, the unit test has failed. With every change that you make to your code, you can run the unit tests and verify that nothing is broken. This gives you confidence in the quality of the code that you've produced.

Unit testing sounds terrific in theory, but setting up unit testing takes significant effort. A big challenge is structuring your code so that your routines are testable outside of the production infrastructure in which they run. This can be straightforward for simple routines that operate only with local variables. But, it is much more complicated for code that requires access to a live SAS environment and data. For the complicated scenarios, most unit testing frameworks support the concept of mock objects. Mock objects enable you to simulate the production infrastructure in which the code runs.

For complex software, unit testing (and, in general, designing for testability) is a proven practice that reduces software defects late in the development cycle. However, the development cycle is front-loaded with tons of work. And, this is when you need to write your unit tests. An alternative is to practice *test-driven design* (or TDD). In TDD, you actually write the tests before you write the software. After the tests are written, you develop the code that makes the tests pass.

The more expensive Team System editions of Microsoft Visual Studio have unit testing support built into them. There are very good open-source unit testing frameworks available at no cost, including NUnit and NMock (for mock objects).

Catch and Handle Exceptions

How should code behave when something unexpected occurs? If you are a SAS programmer, you'll be accustomed to this model:

- An error message or warning is written to the SAS log.
- If the error is severe enough, the current step (DATA step or procedure) stops processing, but SAS still processes subsequent programming statements.
- If the error is caused by a syntax anomaly, such as a missing semicolon or mismatched quotation mark, the state of your SAS session can become unreliable. This means that you cannot safely fix the error, restart the program, and be confident in the results.

In .NET programming, you don't need to worry about syntax errors during run time. The .NET compiler exposes syntax errors before run time. You can't run your program until they are fixed.

However, just because a program compiles doesn't mean that it will run successfully. When an error occurs in a .NET program, an exception is thrown.

Look at the message in Figure 8.2. If you are a user of SAS Enterprise Guide, have you ever seen it before?

Figure 8.2: The Unhandled Exception Message

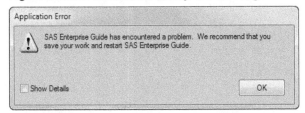

If you have seen it, it's because SAS Enterprise Guide encountered an exception that wasn't properly handled or contained. As a custom task developer, if *you* don't handle or contain the exceptions that your program encounters, your users might see a similar message as they use your task.

Suppose SAS Enterprise Guide is a factory. In a real factory, there are lots of workers organized into a hierarchy, and there is a defined chain of command. In SAS Enterprise Guide, suppose each line of code in your custom task is a worker who has one job—to perform the instruction in the code. When something unexpected happens, the worker cannot recover, so he throws up his hands and tells his supervisor, "Whoa! I cannot complete the job because something unexpected happened! What now?"

The supervisor has two choices. He can handle the unexpected condition (if he knows how), and, as a result, contain the situation. Or, he can throw up his hands and pass the situation on to the next higher level of command.

If no one handles this unexpected condition and it just keeps bubbling up to the top, eventually the factory (or application, as in SAS Enterprise Guide) takes over and shuts down the entire operation (as in your custom task). Your task is terminated and any work done by your task is lost.

Obviously that's not the outcome that you want. And, it doesn't reflect well on the custom task or on the person who developed it. Thus, it's a good idea to prevent exceptions from cascading outside the confines of your task.

Prevent the Preventable Exceptions

The most common exception in .NET programming is the null reference exception. This exception happens when your code tries to access a property or method of a .NET object and the object has no value (that is, it has a null value or references nothing). (It might sound very Zen, but it's frowned upon in programming!)

When this exception happens, the error message looks similar to this:

```
System.NullReferenceException: Object reference not set to an instance
of an object.
```

Even though null reference exceptions are the most common type of exception, they are also the easiest to prevent. It's simple: if there is *any* doubt that an object has a value, check for a null value before you reference it in your code.

Suppose that you have the following line of code:

```
/* if list is null, you can't access Count */
if (list.Count > 0)
{
  /*do work*/
}
```

To prevent a null reference exception, change the code to this:

```
/* if list is null, no attempt to access Count */
if (list!=null && list.Count > 0)
{
  /*do work*/
}
```

If you know for sure that in your program, the value of `list` cannot possibly be null, then prevention isn't necessary. But, if there is *any* possibility of a null value, checking for it is the difference between your task continuing to work or your task being abandoned in the shame of an unhandled exception.

Catch and Handle Recoverable Exceptions

Sometimes your code encounters errors that you cannot avoid. For example, what if your program relies on reading a file, but suddenly the file is removed or locked? Or, what happens when you are accessing a remote data source, but the network connection is suddenly interrupted?

Fortunately, you can teach your program to handle these exceptions. Which conditions you choose to handle depends on the task. For example, can your task continue without a network connection? If it can't, how can you handle this exception?

For each method that you use from a documented API, you should consider whether the method could possibly throw an exception. APIs with good documentation indicate possible exceptions and you can decide whether you want to handle them.

Let's look at the documentation for one of the APIs in SAS.Tasks.Toolkit. The `SasData.GetColumns()` method retrieves a list of `SasColumn` objects (each object represents a variable in a SAS data set). To retrieve that list, the conditions have to be just right. You need a connection to a SAS session, the data must exist, and you must have permission to read the data contents.

Figure 8.3 shows the documentation for the `SasData.GetColumns()` method. In the method documentation, there is a section named "Exceptions" that shows which exceptions this method might trigger and under what conditions.

Figure 8.3: Example of Method Documentation with Exceptions

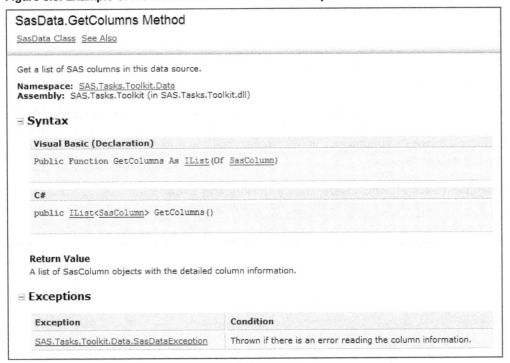

Based on the documentation, a `SasDataException` type of exception might occur if there is a problem while reading the column information. Although this information might not help you prevent the condition that causes the exception, you've at least been warned. It's a condition that's outside of your control. Therefore, you should enclose the `SasData.GetColumns()` method in a Try and Catch block.

A Try and Catch block is a programming construct that enables local exception handling. This is opposed to kicking the exception back up to the calling method or back up all the way to SAS Enterprise Guide. In C#, it looks similar to this:

```
List<SasColumn> cols;
try
{
```

```
    cols = data.GetSasColumns();
}
catch (SAS.Tasks.Toolkit.SasData.SasDataException ex)
{
    /* additional handling if desired */
}
```

In Visual Basic, it looks similar to this:

```
List<SasColumn> cols
Try
    cols = data.GetSasColumns()
Catch ex As SAS.Tasks.Toolkit.SasData.SasDataException
    ' additional handling if desired
End Try
```

With the Try and Catch block, you can take action if the data columns cannot be fetched. You can decide whether this is a condition that you can live with (perhaps you were going to simply display the list of columns for information), or whether this condition is fatal for your task (you cannot continue without data).

Example: Testing for Null Reference and Handling Exceptions

Let's look at an example of how to prevent and handle exceptions.

Consider this snippet of code that uses **SAS.Tasks.Toolkit** to read a saved setting in a custom task:

```
using SAS.Tasks.Toolkit.Helpers;

private void readSavedWidth()
{
    string savedW =
        TaskUserSettings.ReadValue(TaskClassID, "WIDTH");
    int w = Convert.ToInt32(savedW);
    this.Width = w;
}
```

What happens if the **ReadValue()** routine doesn't find a saved value for **WIDTH**? Does it return an empty string? Or, is the value null? The documentation for the API might be able to tell you. Or, you can use the **string.IsNullOrEmpty()** method to test for both conditions. Empty or null, you won't have a value to process, so the subsequent steps are probably going to be the same.

```
private void readSavedWidth()
{
    string savedW =
        TaskUserSettings.ReadValue(TaskClassID, "WIDTH");
    if (!string.IsNullOrEmpty(savedW))
```

```
    {
      int w = Convert.ToInt32(savedW);
      this.Width = w;
    }
  }
}
```

With the method, the code snippet handles both conditions—the string value for **WIDTH** can be null or empty. But, that doesn't mean you can use the value. This code converts the string value into an integer, and then it sets the value for the WIDTH property of a Windows Form. If you look at the documentation for the `Convert.ToInt32()` method in the Microsoft .NET documentation, you see that it might throw exceptions. It throws a `FormatException` if the value isn't numeric or an `OverflowException` if the value is larger than the maximum size for an integer.

For the purpose of this code, you don't need to distinguish between these two exception types. If there is *any* exception, the result is that you won't have a value to set for the WIDTH property. That's all you need to worry about. Therefore, you need to catch *all* exceptions. Setting the Windows Form WIDTH property isn't critical for the task operation, so it's safe to eat this exception and carry on.

```
private void readSavedWidth()
{
  string savedW =
    TaskUserSettings.ReadValue(TaskClassID, "WIDTH");
  if (!string.IsNullOrEmpty(savedW))
  {
    try
    {
      int w = Convert.ToInt32(savedW);
      this.Width = w;
    }
    catch { }
  }
}
```

If you catch all exceptions too broadly (as in the previous example), it can mask real problems. The best practice is to catch as specific an exception as you can. If you cannot be specific, at least limit the scope of code in the Try block.

If you work with a team of developers on shared code, it's a good idea to agree on a strategy for how to handle exceptions. There are entire books devoted to the topic of exception handling, and it can be the source of robust debate among team members.

As a custom task developer, the main thing to remember is this—if your task causes or throws an exception without handling it, two things happen:

- Your task stops processing. Any work done by the task is lost.
- SAS Enterprise Guide catches the exception and displays an unattractive message.

Use Logging to Record Events and Progress

SAS programmers are very familiar with the concept of logs. Every SAS program that you run generates a log that details every instruction that was executed and whether there were any warnings or errors. Understanding the SAS log is fundamental to the success of a SAS programmer.

At the application level, SAS Enterprise Guide has a built-in logging mechanism as well. But, most users never see the contents of an application's log. The log exists primarily for diagnostic purposes. If you call SAS Technical Support with a tricky problem in SAS Enterprise Guide, support personnel might provide instructions on how to collect the log output and send it to SAS for further study. The log is not really customer-serviceable, but it can help the support personnel diagnose a problem that you might be experiencing.

The logging mechanism that SAS Enterprise Guide uses is an open-source standard called log4net. Although it's designed specifically for .NET programs, there are parallel mechanisms for other languages (such as log4j for Java). These logging mechanisms are documented in great detail at http://logging.apache.org. (SAS developed a similar mechanism for enterprise-level logging of SAS components. It is called log4SAS and you can learn more about it by searching for **log4SAS** on http://support.sas.com.)

What Can the Log Tell You?

Logging is a nonintrusive method for recording the state of your application as it runs. And, it is a great debugging method. Other methods of debugging, such as attaching a debugger or showing messages with debugging information, can actually interrupt the normal processing of your application. With these other methods, it is difficult to track down a defect that has a complex navigation path. Or, the observer effect interferes, and you find that the defect doesn't show up when you add a debugger to the process.

The log4net log includes messages with different levels of detail, with each message time-stamped in a predictable way. The messages tell you the state of the application as various parts of your code run. And, the timestamps provide an indication of how long each action took to complete.

Let's look at a sample segment from a log file. This segment has been edited for brevity, but it should be sufficient to give you an idea about how to interpret the log's content.

```
2012-02-19 16:47:20,856 [5] INFO - Workspace request for Local
2012-02-19 16:47:20,857 [5] INFO -    AuthDomain:
2012-02-19 16:47:20,870 [5] INFO - Attempting connect to Local
2012-02-19 16:47:20,870 [5] INFO -    Machine: localhost@
2012-02-19 16:47:20,870 [5] INFO -    LogicalName:
2012-02-19 16:47:20,870 [5] INFO -    IOM options: LOCALE=en-US
2012-02-19 16:47:20,870 [5] INFO -    LoginMode: None
2012-02-19 16:47:20,870 [5] INFO - Login User =, Password = [1]
2012-02-19 16:47:26,115 [5] INFO - Connection successful
```

In this segment of log output, SAS Enterprise Guide is making a connection to a SAS session. The first portion of each line is dedicated to the timestamp. In the timestamp, there is first the date in year-month-day format, and then the time of day in a 24-hour time format. The last three digits represent milliseconds. You can see how precise log timing can be. This entire section of log output spans only six seconds, with most of the messages about the connection that's being made. There is a six-second gap in the messages until the connection is reported as successful.

How to Enable Logging in SAS Enterprise Guide

To turn on the logging facility in SAS Enterprise Guide, you copy a configuration file to the personal profile area where the application loads your preferences and other settings.

A stock configuration file, named logging.config, is provided with SAS Enterprise Guide in the application directory. (The application directory is where SEGuide.exe is located.) When you copy the logging.config file to your personal profile area, logging begins.

The exact location where you copy the logging.config file varies, depending on the version of SAS Enterprise Guide that you are running.

- In SAS Enterprise Guide 4.1, copy logging.config to `%appdata%\SAS\Enterprise Guide\4`. (Note the space in the name.)
- In SAS Enterprise Guide 4.2, copy logging.config to `%appdata%\SAS\EnterpriseGuide\4.2`. (There are no spaces in the name.)
- In SAS Enterprise Guide 4.3, copy logging.config to `%appdata%\SAS\EnterpriseGuide\4.3`. (There are no spaces in the name.)
- In SAS Enterprise Guide 5.1, copy logging.config to `%appdata%\SAS\EnterpriseGuide\5.1`. (There are no spaces in the name.)

Remember, `%appdata%` is a Microsoft Windows environment variable that maps to your personal profile area, which is an area specific to your user account on the PC.

When the logging facility is active, the log files are created in a `Logs` subfolder, which is located near the logging.config file that you've copied. For example, in SAS Enterprise Guide 4.3, the log files are in `%appdata%\SAS\EnterpriseGuide\4.3\Logs`.

Add Your Own Logger and Produce Log Messages

In log4net terminology, a component that emits log messages is a *logger*. In your custom task code, you can create your own logger and produce your own log messages. These messages appear instream with the other log messages that SAS Enterprise Guide creates.

Support for log4net is located in the log4net.dll file. This DLL file is the .NET assembly that is shipped in the SAS Enterprise Guide application directory. Although you might find different versions of log4net available elsewhere, you *must* use the log4net version that is shipped with SAS Enterprise Guide to ensure that there is no version conflict.

To add a reference to log4net in your Visual Studio project:

1. Select **Project→Add Reference**. The Reference Manager dialog box appears.
2. Click the **Browse** button. Navigate to the SAS Enterprise Guide application directory (for example, `C:\Program Files\SASHome\x86\SASEnterpriseGuide\4.3`).
3. Select log4net.dll, and then click **Add**. Click **OK**. The .NET assembly reference is added to the project.
4. In the **References** list in the project solution, click on **log4net**. The log4net properties appear in the Properties view. In the Properties view, change these two settings:
    a. Set **Copy Local** to **False**. (This prevents the DLL file from being copied to the project's build directory.)
    b. Set **Specific Version** to **False**. (This enables you to use this same custom task with future versions of SAS Enterprise Guide even if the log4net.dll version changes.)

To add a logger to your custom task code:

1. Add a `static` class member for the logger and tie it to the object type. For example, if your task class is an object type called SummaryTask, the C# code would look similar to this:

```csharp
using log4net;
static log4net.ILog _logger =
    log4net.LogManager.GetLogger(typeof(SummaryTask));
```

The Visual Basic code would look similar to this:

```vb
Imports log4net
Private _logger As log4net.ILog
_logger = log4net.LogManager.GetLogger(Me.GetType())
```

With the logger defined and initialized, you can add helpful diagnostic messages to the log4net log. For example, suppose your task code encounters an unexpected condition, but you handle it gracefully (from the user's point of view) by catching the exception. You might still want to *log* the exception in case the unexpected condition leads to trouble later on in the process.

Here's an example of handling an exception in C# from earlier in the chapter, but, this time, you log the exception:

```csharp
try
{
   int w = Convert.ToInt32(savedW);
   this.Width = w;
}
catch (Exception ex)
{
   _logger.Error("Could not process saved WIDTH",ex);
}
```

To the user, this behaves the same way it did before. But, if you're trying to diagnose an elusive problem by examining the log4net log, you'll have a helpful hint in the log file:

```
2012-02-22 12:41:16,985 ERROR SummaryTask Could not process
    saved WIDTH  System.Exception.FormatException
...
```

The exception message is accompanied by a detailed stack trace in the log file, which helps reveal exactly where the offending code is lurking.

Use Logging Techniques to Profile Your Task

In addition to logging exceptions, it's a good idea to log operations that might take a long time to run. Some examples of potential bottlenecks include opening data sets, populating UI controls with lists of data, running SAS jobs, and connecting to network resources. By creating log messages when you begin and end these operations, you can later use the log output to determine which operations are consuming the most clock time.

Profiling is the activity of monitoring your application code for performance. The Professional editions of Visual Studio have profiling tools built in, and there are many third-party profiling tools that can help .NET application developers. The process of profiling usually readies your code for reporting (by injecting timers and counters).

The main insights that profiling provides are:

- How long does each instruction and routine take to execute?
- How many times is each instruction or routine called?

You can use profiling to provide this insight about your custom tasks. Or, you can use logging to provide some of the same insight with less process and cost.

Suppose you have a .NET routine that retrieves the properties of a SAS variable. (This is a useful operation in a custom task.) Because this operation requires connection to a SAS session and reads data, it is subject to delays associated with data characteristics, server performance, and network latency. In anticipation of these potential bottlenecks, you can add logging to your code:

```
using SAS.Tasks.Toolkit;
using SAS.Tasks.Toolkit.Data;
using log4net;

/* inside one of your task routines */
string server = "Local",
    lib = "SASHELP",
    mem = "CARS";
log4net.ILog logger = log4net.LogManager.GetLogger("DataLogger");
logger.InfoFormat("Accessing columns for {0}.{1}", lib, mem);
SasData data = new
```

```
    SAS.Tasks.Toolkit.Data.SasData(server,lib,mem);
foreach (SasColumn c in data.GetColumns())
{
    logger.InfoFormat("Retrieving properties for column {0}", c.Name);
}
logger.InfoFormat("COMPLETE: Accessed columns for {0}.{1}", lib, mem);
```

When you run your task in SAS Enterprise Guide, and then later examine the log4net log output, you find blocks of messages similar to this:

```
2012-02-25 11:00:49,606 INFO   DataLogger - Accessing columns for
SASHELP.CARS
2012-02-25 11:00:50,743 INFO   DataLogger - Get properties for Make
2012-02-25 11:00:50,743 INFO   DataLogger - Get properties for Model
2012-02-25 11:00:50,743 INFO   DataLogger - Get properties for Type
2012-02-25 11:00:50,743 INFO   DataLogger - Get properties for Origin
2012-02-25 11:00:50,743 INFO   DataLogger - Get properties for
DriveTrain
2012-02-25 11:00:50,743 INFO   DataLogger - Get properties for MSRP
2012-02-25 11:00:50,743 INFO   DataLogger - Get properties for Invoice
2012-02-25 11:00:50,743 INFO   DataLogger - Get properties for
EngineSize
2012-02-25 11:00:50,743 INFO   DataLogger - Get properties for
Cylinders
2012-02-25 11:00:50,743 INFO   DataLogger - Get properties for
Horsepower
2012-02-25 11:00:50,743 INFO   DataLogger - Get properties for MPG_City
2012-02-25 11:00:50,743 INFO   DataLogger - Get properties for
MPG_Highway
2012-02-25 11:00:50,743 INFO   DataLogger - Get properties for Weight
2012-02-25 11:00:50,743 INFO   DataLogger - Get properties for
Wheelbase
2012-02-25 11:00:50,743 INFO   DataLogger - Get properties for Length
2012-02-25 11:00:50,743 INFO   DataLogger - COMPLETE: Accessed columns
for SASHELP.CARS
```

Reviewing this log output provides some insight.

- In this case, the initial access operation for the data set took just about one second (from 11:00:49 to 11:00:50).
- The time to access each individual column property was negligible.

This is good information that might help you as you develop during the testing phase. However, suppose that after task development is completed and deployed to a field of users, some users complain about slow performance in the task.

You can tell users how to enable log4net logging, have them work as usual, and then have them send you the log file. The logging output might reveal that for the data the user is accessing, it takes

10 seconds to access the column metadata. What might cause such a delay? Network performance? It might be that the data member is a data view (which takes longer to render), not a SAS data set. Or, it might be based on how the user is interacting with the task (the user is accessing the data multiple times). Unlike the reports from users, log4net log output provides objective answers to the questions about what's happening in your application. The log4net log is an important forensic tool for helping you find the root cause of the users' complaints.

Debug with Microsoft Visual Studio

The Professional editions of Visual Studio include a powerful debugger that you can use to stop at breakpoints and to examine the state of your custom task code while it is running. The debugger enables you to peek at your task's methods, properties, and variables, and to walk you through your code one step at a time to watch the effect of your .NET language statements.

Debugging Basics: Some Definitions

Before learning the techniques for debugging a custom task, it's important to understand a few basic debugging terms. Here are some terms that you need to know:

.NET assembly (DLL file)
The .NET assembly, also known as a DLL file, is the file that contains your compiled code. It's usually the only file that you need to deploy to a user's machine when your custom task is completed.

Source project or solution (csproj, vbproj, or sln file)
The source project is the .NET source code (C# or Visual Basic) that you write and maintain for your custom task. The code source files are either C# (.cs) files or Visual Basic (.vb) files. When you run the debugger, you usually have the source project open in Visual Studio so that you can navigate the code source files, add breakpoints, and examine the values of local variables.

Symbols file (PDB file)
The symbols file is created when you build your .NET source project. It contains information about your source project that helps the debugger connect your compiled code (in the DLL file) to the original code source files and statements. The symbols file serves as a map from the compiled code to the code source files and statements. Without the symbols file, you cannot effectively use the debugger to set breakpoints.

Breakpoint
A breakpoint is a point at which code execution stops and control is handed over to the debugger. You can set breakpoints by marking statements in your code source file or by defining conditions for triggering a breakpoint (for example, an unhandled exception). When a breakpoint is encountered, code execution is paused, and you can use the debugger to examine the state of your code.

Prepare to Debug a Custom Task

Before you can effectively debug a custom task (or any application), you need to make sure that the debugger can locate the symbols for the modules that you want to debug. Without the symbols (which are located in the PDB file), you won't be able to view the source code of the content that you are debugging, and you won't be able to set effective breakpoints.

With custom tasks, it's important to remember how a SAS application loads the custom task modules. Loading is described in detail in "Deploying Custom Tasks," in Chapter 1, "Why Custom Tasks?" For effective debugging, you need to make sure that the associated PDB files are in the same location as the custom task DLL files that are loaded.

How to Attach a Debugger to a Custom Task

To use the Visual Studio debugger, you must attach the debugger to an application or to a Windows process. Once a debugger is attached to an application, it monitors the application instructions and respects debugger directives such as "stop at my breakpoints." Keep in mind that a custom task is *not* a stand-alone application. It behaves as a plug-in to another application that serves as its host, such as SAS Enterprise Guide or Microsoft Office (when used with the SAS Add-In for Microsoft Office).

For SAS Enterprise Guide, the process name is `SEGuide.exe`. For the SAS Add-In for Microsoft Office, the process name corresponds to the Microsoft Office application that you are using (for example, `EXCEL.exe`, `WINWORD.exe`, and `POWERPNT.exe` for Microsoft Excel, Word, and PowerPoint, respectively).

There are two main methods for attaching a debugger to an application or process:

1. Launch the application from the debugger. With this method, you create a Windows command that begins your debugging session, and that same command launches the application that you want to debug. The application begins with the debugger attached right from the start, and the debugger recognizes any breakpoints that occur early in the application.
2. Attach the debugger to a running process. With this method, the debugger attaches mid-stream to a process that is already running on your computer. Once it is attached, the debugger recognizes any breakpoints that occur from that point on.

Each method has its advantages. The first method offers convenience because you can change the code in your task, rebuild it, and then launch the application from the debugger. You catch early breakpoints because the debugger is monitoring the application and your task behavior from the very beginning.

However, using a debugger introduces overhead that can cause the application to perform slower. To help things run a little bit faster, in method 2, you can attach the debugger to the application right before the point that you want to observe, and you can then stop at the breakpoints that you are interested in. This method requires more familiarity with the sequence of events (for example, when your task is loaded, initialized, and run), but it can save you a little bit of time.

Note for Users of Express Editions If you are using one of the Express editions of Visual Studio, using the debugger with either of the previous methods presents a special challenge. That's because the Express editions do not allow you to launch the debugger with a start-up command for an external application. You can debug only applications that you create in the Express edition environment. Because your custom task is a plug-in, not an application, you cannot launch the debugger in an Express edition. Also, the Express editions do not allow you to attach the debugger to a running process. This means that Express edition users cannot use either of the previous methods for attaching the debugger.

If you are using an Express edition, what recourse do you have? Obviously, Microsoft wants to encourage serious application developers to invest in a license for a Professional edition. But, if you want to continue with the free Express editions, you have to endure these debugging inconveniences.

There are a few workarounds, but the inconvenience might be enough to nudge you toward purchasing a license for a Professional edition. One workaround is to create an application that does nothing but launch the real application that you want to debug. You then launch the first application from the debugger. Another workaround involves an undocumented method for tricking the Express edition of Visual Studio to launch the application that you want to run. Because both of these workarounds are subject to change and because they run contrary to the purpose of the Express editions, details are not included in this book.

Example: Debugging a Custom Task

The projects used in the following example are available at http://go.sas.com/customtasksapi. However, the techniques are generic enough to apply to any custom task project.

In this example, you open a Visual Studio solution, select which project to debug, define the start-up command, set breakpoints, and launch the debugger.

1. Open Visual Studio.
2. Select **File→Open Project.** Navigate to the Examples.sln project. Select it, and then click **Open**. The Solution Explorer view should look similar to Figure 8.4.
3. In the Solution Explorer view, select **SASProgramRunner** and right-click on it. Select **Set as StartUp Project**. This makes this project the primary project when you launch the debugger.
4. Right-click on **SASProgramRunner** again, and select **Properties**. The project properties appear. The properties have several tabs that represent different aspects of the project, such as **Application**, **Compile**, **Debug**, **Settings**, and more.
5. Select the **Debug** tab. The contents should look similar to Figure 8.5.
6. In the **Start Action** section, select **Start external program**. In the text field, enter the full path for SAS Enterprise Guide. You can use the browse button (**…**) to navigate to **SEGuide.exe**, or you can enter the full path yourself. For SAS Enterprise Guide 4.3, the full

path is similar to `C:\Program Files\SAS\EnterpriseGuide\4.3\SEGuide.exe`.

Figure 8.4: The Solution Explorer View

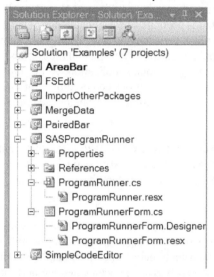

Figure 8.5: The Debug Tab of Properties

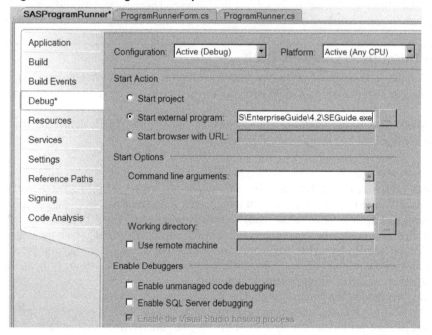

To verify that the debug settings are correct, you can set a breakpoint and launch the application. For example, suppose you want to set a breakpoint that is triggered when you click the **Run program** button of the **SASProgramRunner** project. Here are the steps:

1. In the **SASProgramRunner** project, right-click on the **ProgramRunnerForm.cs** file, and select **View Code**. The C# source file opens in the code editor view.
2. Locate the event handler code that is triggered by clicking the button. In this case, the function includes `btnSubmit_Click`.
3. In the first line of that function, click in the left margin of the code editor to set a breakpoint. The breakpoint is indicated by a round dot in the left margin. The affected code is highlighted. The code editor should look similar to Figure 8.6.
4. To launch the debugger, select **Debug→Start Debugging**. (A shortcut for this action is the F5 key.)

Figure 8.6: A C# Source File with a Breakpoint Set

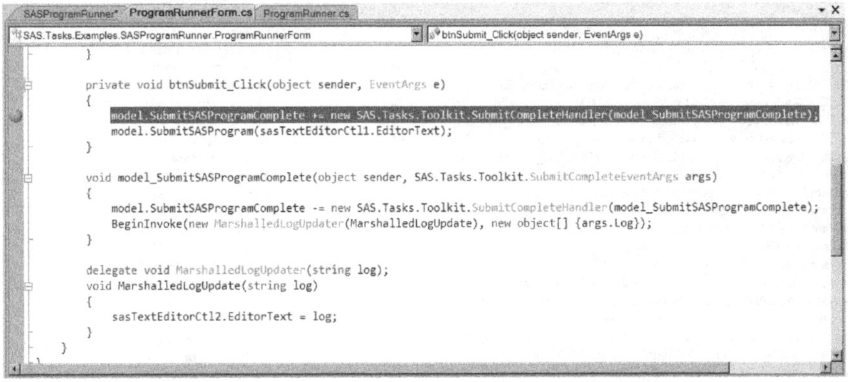

SAS Enterprise Guide is launched as the host application. Select **Tools→Add-In→SAS Examples→SAS Program Runner** to launch the task. When the SAS Program Runner dialog box appears, click **Run program**.

If the breakpoint was set correctly, the focus will shift back to Visual Studio, and the debugger is positioned at the breakpoint that you set. You can use the debugger to examine the states of variables in the task, step through the code, and control the flow of execution.

Troubleshooting: When Breakpoints Aren't Hit

When your breakpoints aren't hit when you expected, the problem is one of two things:

1. The PDB file was not in the same location as the custom task DLL file that was loaded. You can fix this by ensuring that the PDB file that was created from your custom task build was copied, along with your DLL file, to the `Custom` task folder that you are using.
2. The custom task DLL file was loaded from a location different from what you expected. Because SAS applications offer different options for locating custom task DLL files, it is

possible that a copy of your custom task DLL file was loaded from a different location, and that the DLL file that was loaded is not the version that you want to debug.

A quick way to diagnose the second problem is to examine the process and verify that the DLL file was loaded from the location that you expected. The Visual Studio debugger has a built-in tool that makes this diagnosis easy.

With the debugger attached and running in SAS Enterprise Guide, in Visual Studio, select **Debug→Windows→Modules**. Figure 8.7 shows an example of the Modules view in a debugging session. You can examine the **Name** and **Path** values in the list to verify that each module is loaded from the location that you expected, and that the debugger recognizes the symbols (from the PDB file).

Figure 8.7: The Modules View of the Debugger

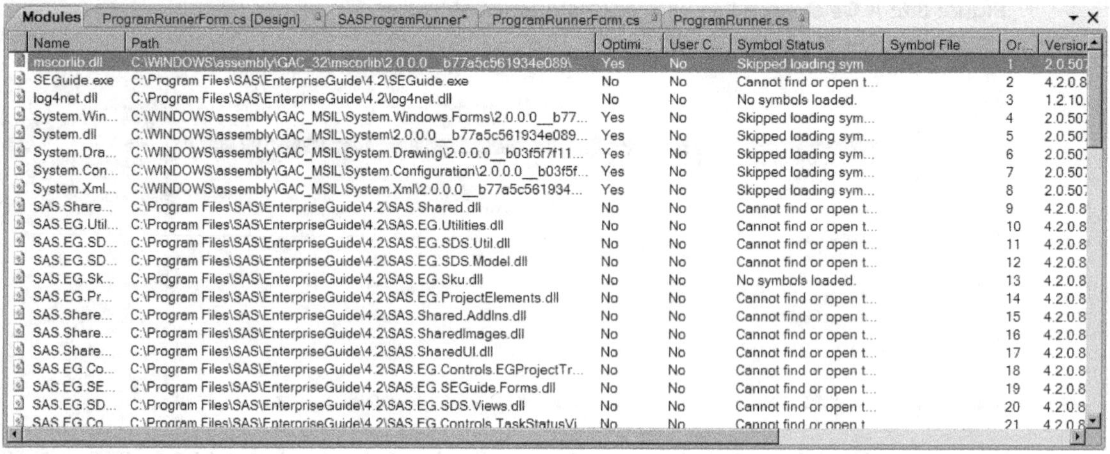

Name	Path	Optimi...	User C...	Symbol Status	Symbol File	Or...	Version
mscorlib.dll	C:\WINDOWS\assembly\GAC_32\mscorlib\2.0.0.0__b77a5c561934e089\	Yes	No	Skipped loading sym...		1	2.0.50
SEGuide.exe	C:\Program Files\SAS\EnterpriseGuide\4.2\SEGuide.exe	No	No	Cannot find or open t...		2	4.2.0.8
log4net.dll	C:\Program Files\SAS\EnterpriseGuide\4.2\log4net.dll	No	No	No symbols loaded.		3	1.2.10.
System.Win...	C:\WINDOWS\assembly\GAC_MSIL\System.Windows.Forms\2.0.0.0__b77...	Yes	No	Skipped loading sym...		4	2.0.50
System.dll	C:\WINDOWS\assembly\GAC_MSIL\System\2.0.0.0__b77a5c561934e089...	Yes	No	Skipped loading sym...		5	2.0.50
System.Dra...	C:\WINDOWS\assembly\GAC_MSIL\System.Drawing\2.0.0.0__b03f5f7f11...	Yes	No	Skipped loading sym...		6	2.0.50
System.Con...	C:\WINDOWS\assembly\GAC_MSIL\System.Configuration\2.0.0.0__b03f5f...	Yes	No	Skipped loading sym...		7	2.0.50
System.Xml...	C:\WINDOWS\assembly\GAC_MSIL\System.Xml\2.0.0.0__b77a5c561934...	Yes	No	Skipped loading sym...		8	2.0.50
SAS.Share...	C:\Program Files\SAS\EnterpriseGuide\4.2\SAS.Shared.dll	No	No	Cannot find or open t...		9	4.2.0.8
SAS.EG.Util...	C:\Program Files\SAS\EnterpriseGuide\4.2\SAS.EG.Utilities.dll	No	No	Cannot find or open t...		10	4.2.0.8
SAS.EG.SD...	C:\Program Files\SAS\EnterpriseGuide\4.2\SAS.EG.SDS.Util.dll	No	No	Cannot find or open t...		11	4.2.0.8
SAS.EG.SD...	C:\Program Files\SAS\EnterpriseGuide\4.2\SAS.EG.SDS.Model.dll	No	No	Cannot find or open t...		12	4.2.0.8
SAS.EG.Sk...	C:\Program Files\SAS\EnterpriseGuide\4.2\SAS.EG.Sku.dll	No	No	No symbols loaded.		13	4.2.0.8
SAS.EG.Pr...	C:\Program Files\SAS\EnterpriseGuide\4.2\SAS.EG.ProjectElements.dll	No	No	Cannot find or open t...		14	4.2.0.8
SAS.Share...	C:\Program Files\SAS\EnterpriseGuide\4.2\SAS.Shared.AddIns.dll	No	No	Cannot find or open t...		15	4.2.0.8
SAS.Share...	C:\Program Files\SAS\EnterpriseGuide\4.2\SAS.SharedImages.dll	No	No	Cannot find or open t...		16	4.2.0.8
SAS.Share...	C:\Program Files\SAS\EnterpriseGuide\4.2\SAS.SharedUI.dll	No	No	Cannot find or open t...		17	4.2.0.8
SAS.EG.Co...	C:\Program Files\SAS\EnterpriseGuide\4.2\SAS.EG.Controls.EGProjectTr...	No	No	Cannot find or open t...		18	4.2.0.8
SAS.EG.SE...	C:\Program Files\SAS\EnterpriseGuide\4.2\SAS.EG.SEGuide.Forms.dll	No	No	Cannot find or open t...		19	4.2.0.8
SAS.EG.SD...	C:\Program Files\SAS\EnterpriseGuide\4.2\SAS.EG.SDS.Views.dll	No	No	Cannot find or open t...		20	4.2.0.8
SAS.EG.Co...	C:\Program Files\SAS\EnterpriseGuide\4.2\SAS.EG.Controls.TaskStatusVi...	No	No	Cannot find or open t...		21	4.2.0.8

Chapter Summary

In a custom task project or in any application for that matter, you might find that you spend as much time chasing down bugs as you do developing the program code in the first place. By following good design techniques, responsible exception handling, and prolific logging, you can improve the quality of your code and reduce the amount of time that you spend debugging.

Chapter 9: The Top N Report

About This Example	**118**
Example Source Files and Information	118
Step 1: Exploring the Problem	**119**
Step 2: Creating the SAS Program	**121**
Step 3: Creating the Custom Task	**123**
Examining the Top N Report Solution	**123**
Chapter Summary	**129**

The Top N report is pervasive in our society. From the *Billboard* Hot 100 to *The New York Times* Best Seller list to the *Forbes* 400 Richest People in America, the Top N report is a medium that everyone understands. It answers the question, "Which are the top that stand out from the rest?"

Businesses rely on reports like the Top N every day. It begins with what seem like simple questions.

- Who are our top 10 customers, measured by amount of sales?
- Who are the 10 most effective sales associates in each region?
- Which are the five top-selling brands for each product family that we distribute?

There are many different ways in SAS to create reports that answer these types of questions. If you support business users of corporate data, your job is to provide a consistent way to answer these questions in a format that is easy to understand and use.

About This Example

The task example illustrates several techniques and patterns that you can apply in your own custom task, including:

- A reusable SAS program that uses SAS macro variables to support a variety of scenarios.
- An elegant user interface that reads and displays the properties of an input data set, including column names, types, and formats.
- A simple pattern that you can follow for storing user selections (called *serialization*) so that they are remembered as part of your SAS Enterprise Guide project.

Example Source Files and Information

The task example is named **SASPress.CustomTasks.TopN.dll**. It is built and ready to use. Here are some details about the task example:

.NET language and version	C# and .NET 1.1 (Microsoft Visual Studio 2003 and later)
.NET difficulty	Low
.NET features	Windows Forms
SAS difficulty	Low
SAS features	MEANS, REPORT, and SQL procedures
Binaries	SASPress.CustomTasks.TopN.zip
Source code	SASPress.CustomTasks,TopN_src.zip

The source code is in the Microsoft Visual Studio solution named **TopN_CSharp.sln**. For more information about how to access source code, see "Accessing Ready-to-Use Example Tasks and Source Code," in Chapter 1, "Why Custom Tasks?"

Step 1: Exploring the Problem

The first step to designing an effective task is to make sure that you understand the problem domain. And, the best way to make sure that you have a correct understanding is to talk to the users who will eventually use the task's reports. To help facilitate this communication, it's often best to begin with an answer to the question your users are asking. Create a few prototype reports that represent what you can create from this task and make sure everyone agrees on what the solution should be. Here are some questions that you should be able to answer before you invest in implementing the ultimate solution:

- What do you want the results to look like? In addition to a table of results, would a chart help the user interpret the data?
- What type of SAS program would provide the results that you're looking for?
- How will the SAS program need to change to accommodate different data sources and different forms of the question?
- What control does the user need over the input and the results?

Figure 9.1 shows an example of how the report might answer the question, "Which 10 models of automobiles feature the most powerful engines as measured by horsepower?"

Figure 9.1: An Example Report of the Top Automobile Models

Most powerful models by horsepower

Rank	Model	Horsepower
1	G35 4dr	520
2	Viper SRT-10 convertible 2dr	500
3	CL600 2dr	493
4	SL55 AMG 2dr	493
5	SL600 convertible 2dr	493
6	911 GT2 2dr	477
7	RS 6 4dr	450
8	C320 4dr	430
9	Phaeton W12 4dr	420
10	S-Type R 4dr	390

For the answer to a slight variation of the question, you might be interested in the same ranking of automobiles, but grouped by their countries of origin. Figure 9.2 shows an example of what that report might look like.

Figure 9.2: An Example Report of the Top Automobile Models Grouped by Country of Origin

Most powerful models by horsepower, across countries

Origin	Rank	Model	Horsepower
Asia	1	G35 4dr	520
	2	M45 4dr	340
	3	Q45 Luxury 4dr	340
	4	Land Cruiser	325
	5	FX45	315
	6	Pathfinder Armada SE	305
	7	Titan King Cab XE	305
	8	GS 430 4dr	300
	9	Impreza WRX STi 4dr	300
	10	SC 430 convertible 2dr	300
Europe	1	CL600 2dr	493
	2	SL55 AMG 2dr	493
	3	SL600 convertible 2dr	493
	4	911 GT2 2dr	477
	5	RS 6 4dr	450
	6	C320 4dr	430
	7	Phaeton W12 4dr	420
	8	S-Type R 4dr	390
	9	XJR 4dr	390
	10	XKR convertible 2dr	390
USA	1	Viper SRT-10 convertible 2dr	500
	2	Corvette 2dr	350
	3	Corvette convertible 2dr	350
	4	Escalade EXT	345
	5	GTO 2dr	340
	6	Yukon XL 2500 SLT	325
	7	SRX V8	320
	8	XLR convertible 2dr	320
	9	H2	316
	10	Excursion 6.8 XLT	310

After you and your users have agreed on the report specifications and inputs, you can sketch out a user interface. Figure 9.3 shows an example Windows Form for the user interface, which is meant to provide a balance of simplicity and flexibility.

Figure 9.3: The Example User Interface Windows Form for the Top N Report Task

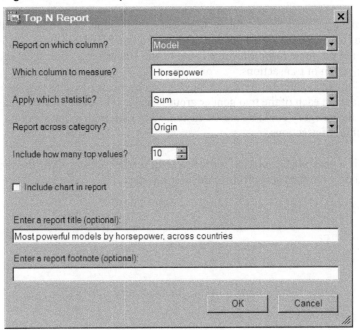

Each of the first four fields provides a list of valid selections for the data. To ensure valid inputs, the task reads the input data characteristics and populates the picklists accordingly. Later in this chapter, you will learn a technique for doing this.

The remaining fields—how many top values to include, whether to include a chart, and what the title and footnote text should say—are user-supplied values that require less validation.

Step 2: Creating the SAS Program

SAS offers many procedures to create reports like these. This task example uses the following steps:

- Use the MEANS procedure to summarize data, either from the entire data set or from data from a grouping variable using the CLASS statement.
- Use the SORT procedure to rank the summarized data, and then use the DATA step to include only the top *n* values in the results.
- Use the REPORT procedure to create an easy-to-read report.
- Use the GCHART procedure to create a simple bar chart that communicates the results in a visual way.

This example supports two main scenarios:

1. A straight report of the top values for a column, aggregated for multiple occurrences of the same response value (for example, the 10 patients with the highest blood pressure). Or, multiple occurrences of the same response value could be the 10 customers that spent the most money in a collection of transactions.
2. A stratified report of the top values for a column, aggregated by a category that you select (for example, the 10 patients in each of the two gender groups (male and female) with the highest blood pressure). Or, the category could be the 10 customers in each region that spent the most money in a collection of transactions.

The following SAS program is a generalized solution for the straight report scenario. Only the first nine lines would differ, based on what the user selects in the user interface.

```
%let data=SASHELP.CARS;
%let report=Model;
%let measure=Horsepower;
%let measureformat=;
%let stat=SUM;
%let n=10;
title "Most powerful models by horsepower, across countries";
footnote;
%let category=Origin;
/* summarize the data across a category and store */
/* the output in an output data set */
proc means data=&data &stat noprint;
     var &measure;
     class &category &report;
     output out=summary &stat=&measure &category /levels;
run;

/* store the value of the measure for ALL rows and
/* the row count into a macro variable for use    */
/* later in the report */
proc sql noprint;
select &measure,_FREQ_ into :overall,:numobs
from summary where _TYPE_=0;
select count(distinct &category) into
:categorycount from summary;
quit;

/* sort the results so that you get the TOP values */
/* rising to the top of the data set */
proc sort data=work.summary out=work.topn;
  where _type_>2;
  by &category descending &measure;
run;

/* Pass through the data and output the first N */
```

```
  /* values for each category */
data topn;
  length rank 8;
  label rank="Rank";
  set topn;
  by &category descending &measure;
  if first.&category then rank=0;
  rank+1;
  if rank le &n then output;
run;

/* Create a report listing for the top values */
/* in each category */
footnote2 "&stat of &measure for ALL values of &report:
&overall (&numobs total rows)";
proc report data=topn;
      columns &category rank &report &measure;
      define &category /group;
      define rank /display;
      define &measure / analysis &measureformat;
run;
quit;
```

Step 3: Creating the Custom Task

Believe it or not, once you have the problem domain defined and the SAS program written, the hard part of the task is done! You have completed the meat of the task—crafting the part of the task that creates the results. The remainder of the task is simple plumbing to hook up your solution to a user interface that enables a user to use it.

You can create the shell of the custom task by following the steps in Chapter 3, "Creating Custom Task Projects in Microsoft Visual Studio." The remainder of this chapter examines some of the techniques used to create the functionality of the task.

Examining the Top N Report Solution

The source of the Microsoft Visual Studio solution for the Top N Report task represents the typical structure of a custom task.

The C#-based solution contains the following main components:

Programs folder with SAS programs
The **Programs** folder contains four SAS programs to support the different reports. StraightReport.sas and StraightChart.sas provide the basis for the basic Top N report and chart with no category column. StratifiedReport.sas and StratifiedChart.sas support a category

column. The processing is slightly different between the types. It is cleaner to keep the implementations separate rather than complicate a single SAS program to support all implementations.

TopNReport class (TopNReport.cs)
The TopNReport class implements the main ISASTask interfaces and enables the task to plug in to SAS Enterprise Guide and the SAS Add-In for Microsoft Office. This class generates the SAS program (using the files in the `Programs` folder) based on the user-supplied values. The TopNReport class models the business logic for creating the report and storing the user-supplied values. As a result, this type of class is sometimes called the *task model*.

TopNReportForm class (TopNReportForm.cs)
The TopNReportForm class is a Windows Form that provides the user interface for the task. Its main purpose is to make it simple for the user to select columns for and control the behavior of the report. This type of class is often called the *task view*.

TopNSettings class (TopNSettings.cs)
The TopNSettings class represents the task's state and provides an abstract layer of persistence that makes it easier to remember the task settings between uses.

All of the source files in the task are well documented with comments in the source code. The remainder of this section provides more details about techniques that the example uses to support important features of its behavior.

Technique: Read and Show the Available Columns

In this task example, it's important to show the user what data columns are available to report on. To determine this, you need to crack open the selected input data just long enough to peek at the available columns and their attributes. Once you have that information, you can populate the user interface controls (such as the combo boxes) with appropriate values.

Figure 9.4 is an example of a combo box that is populated with available columns:

Figure 9.4: Available Columns, Ripe for the Picking

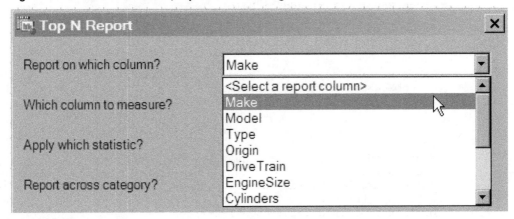

The following C# routine, which is part of the TopNReportForm class (located in TopNReportForm.cs), reads the data and initializes the combo box:

```
const string noCategory = "<No category>";  ❶
const string selectReportPrompt = "<Select a report column>";
const string selectMeasurePrompt = "<Select a measure>";
System.Collections.Hashtable hashFormats = new Hashtable();
private void LoadColumns()
{
  cmbReport.Items.Add(selectReportPrompt);
  cmbMeasure.Items.Add(selectMeasurePrompt);

  try
  {
    // to populate the combo boxes with the available columns,
    // you have to examine the active data source.
    ISASTaskData data = Model.Consumer.ActiveData;
    ISASTaskDataAccessor da = data.Accessor;
    // to do that, you need to "open" the data to get a
    // peek at the column
    // information
    if (da.Open()) ❷
    {
      for (int i=0; i<da.ColumnCount; i++)
      {
        // first add the most likely categories  ❸
        // -- character and date columns
        ISASTaskDataColumn ci = da.ColumnInfoByIndex(i);
        if (ci.Group == VariableGroup.Character ||
            ci.Group == VariableGroup.Date)
```

```
        {
          cmbReport.Items.Add(ci.Name);
          cmbCategory.Items.Add(ci.Name);
        }
        else
        {
          // numeric columns are added to the Measures list
          cmbMeasure.Items.Add(ci.Name);
          hashFormats.Add(ci.Name, ci.Format);
        }
      }

      // now, add the rest of the numerics for Report and
      // Category.
      // These are less likely to make sense, but you don't want
      // to shut the door on "creative" reports.
      for (int i=0; i<da.ColumnCount; i++)
      {
        ISASTaskDataColumn ci = da.ColumnInfoByIndex(i);
        if (ci.Group == VariableGroup.Numeric) ❹
        {
          cmbReport.Items.Add(ci.Name);
          cmbCategory.Items.Add(ci.Name);
        }
      }

      cmbCategory.Items.Add(noCategory);

      da.Close();❺
    }
    else
    {
      // something went wrong in trying to read the data, so
      // report an error message and end the form.
      string dataname = "UNKNOWN";
      if (Model.Consumer.ActiveData!=null) ❻
        dataname = string.Format("{0}.{1}",
          Model.Consumer.ActiveData.Library,
          Model.Consumer.ActiveData.Member);
      MessageBox.Show(
        string.Format("Could not read columns from data {0}.",
          dataname));
      this.Close();
    }
  }
  catch (Exception ex)
```

```
      {
        // log exception
        string msg =
         string.Format("ERROR: Could not load data columns. ({0})",
            ex.Message);
        MessageBox.Show(msg);
      }
    }
```

❶ String constants are used for stock text, such as **Select a report column** and **No category**.

❷ Use the ISASTaskDataAccessor interface, provided by SAS Enterprise Guide in the ISASTaskConsumer interface, to open the data to read the column information.

❸ Iterate through the columns and examine the types for each column. Character and date columns make the most sense to use as report or category columns, so add those types to the list first. In this example, SAS format information about each column is stored. This will come in handy later when you want to generate a SAS program that preserves the format of the column that you are measuring. For example, you'll want to preserve a currency format (DOLLARw.) for a column that represents a value in dollars.

❹ Take a second pass through the columns and append any remaining numeric columns to the report and category lists. Although it is not typical, it might be appropriate to use a numeric column as a category. For example, suppose that there is a numeric column (that is not formatted as a date variable) that represents the year—that's clearly a category, not a measure.

❺ If you open the data with an `Open` call, always make sure that you match it with a corresponding `Close` call when you are finished.

❻ It's always a good idea to anticipate what might go wrong and add some error-handling code. An operation such as reading data requires favorable conditions to succeed. If anything goes wrong, it's better to handle errors and exceptions rather than to confuse the user with unhandled errors and exceptions.

Technique: Save the Task Settings Using a Serializable Class

For your task to produce repeatable results, you must be able to save the task's state—that is, you must be able to preserve the user's task settings between uses of the task. In this example, you need to save the names of the columns, whether to create a chart, and what the title and footnote should say (if anything).

The custom task infrastructure assumes that you are using XML as the format to persist your task's state. If you want to have complete control over the structure of the XML, you can implement code to read and write this XML explicitly, using built-in .NET classes that can read and write XML documents.

However, if you want to simply encapsulate your task's state in a .NET class, you can use built-in XML serialization support to store and retrieve your task's state implicitly with just a few lines of code.

In this task example, the TopNSettings class (located in TopNSettings.cs) stores all of the task settings. The class implementation is tagged as serializable using a .NET class attribute. Here is a snippet of the class declaration:

```
/// <summary>
/// A serializable class that represents the
/// settings for the Top N Report.
/// </summary>
[Serializable]
public class TopNSettings
{
  /// <summary>
  /// a member variable that tracks the
  /// report column names
  /// </summary>
  private string _reportColumn="";
  /// <summary>
  /// The property used by other classes to
  /// set/get the report column name.
  /// </summary>
  public string ReportColumn
  {
    get { return _reportColumn; }
    set { _reportColumn = value; }
  }
  private string _categoryColumn="";
  /// <summary>
  /// The property used by other classes to
  /// set/get the category column name.
  /// </summary>
  public string CategoryColumn
  {
    get { return _categoryColumn; }
    set { _categoryColumn = value; }
  }
// ... remainder of class omitted
```

The `[Serializable]` tag tells the .NET run time that all of the public properties of the TopNSettings class should be saved when the class is persisted. You can use the XmlSerializer class to trigger-transform an instance of the TopNSettings class to XML, and then later convert it from XML to a new instance of the class.

Here is an example from the WriteXML() method in the TopNReport class. It fills an instance of the TopNSettings class with the task settings, and then saves it to XML.

```
TopNSettings settings = new TopNSettings();
settings.CategoryColumn = CategoryColumn;
settings.IncludeChart = IncludeChart;
settings.MeasureColumn = MeasureColumn;
```

```
settings.N = N;
settings.ReportColumn = ReportColumn;
settings.Statistic = Statistic;
settings.Title = Title;
settings.Footnote = Footnote;
settings.MeasureFormat = MeasureFormat;

using (StringWriter sw = new StringWriter())
{
  XmlSerializer s = new XmlSerializer(typeof(TopNSettings));
  s.Serialize(sw, settings);
  return sw.ToString();
}
```

The result is a character string of XML, which SAS Enterprise Guide stores in the project. Later, when the task is rerun, the ReadXML() method in the TopNReport class is used to rehydrate the task's state from these saved task settings.

Note: Using the XmlSerializer class can be a convenient way to save and restore your task state, but it does have some drawbacks.

- The XmlSerializer class relies on a .NET feature called *reflection*, which uses a .NET class definition to dynamically create .NET run-time code to make the serialization work. The reflection feature can perform slowly for large classes or classes with many dependencies on other classes. For simple classes such as the one in this example, the performance impact is negligible.
- Using the XmlSerializer class provides a layer of abstraction so that you don't have to work with the raw XML format. However, the more familiar you are with the XML format, the better you will be able to support your task in the long term if it evolves into many new versions. If the task settings that you save are likely to change in later versions of your task, you might consider reading and writing the XML directly so that future versions of your task can handle task state information created by previous versions.

Chapter Summary

The Top N Report example is an example of a classic task workflow. It provides a user interface to collect information from a user, and then generates a SAS program based on the user's selections.

This task example shows how you can examine and interpret data attributes to help present useful information to the user, and then generate results that match the user's requests.

A solid task implementation is important, but it requires research of the problem space. This research involves discovering what the users want to achieve, prototyping the results using traditional SAS programming, and mocking up a user interface that your users can validate.

Chapter 10: For the Workbench: A SAS Task Property Viewer

About This Example **131**
    Example Source Files and Information 132
What's in Your Project? **132**
    Displaying Properties in a Simple User Interface 133
    Accessing Properties Using the ISASProject APIs 134
    More Possibilities with SAS Enterprise Guide Projects 138
Chapter Summary **139**

Crafting software is similar to any other craft—to succeed in the craft in the long term, you need proper tools.

There are tools that you can buy off the shelf, but they get you only so far. When you have a custom project with unique requirements, you often need to *build* additional tools to support your craft before you can proceed with the custom project.

If you have ever practiced or studied the craft of woodworking, you know that in the beginning, the first projects are not glamorous pieces of furniture or saleable sculptures that would fetch top dollar at a craft show. Your first projects are jigs, box fences, or other items that no one else will ever see, but that are necessary to prepare you for the construction of your final projects.

This chapter shows you how to build a custom task that you will find useful as you develop other custom tasks. It's called the SAS Task Property Viewer, and it enables you to view the attributes of other tasks that are already saved in your project.

About This Example

The task example uses new APIs that were added in SAS Enterprise Guide 4.2. These APIs enable the active task to see what elements are in the active project files. Because this task example is specific to SAS Enterprise Guide project files, it does not work in the SAS Add-In for Microsoft Office.

An overview of these APIs is included in Chapter 4, "Meet the Custom Task APIs." Specifically, the APIs that enable you to see inside a project are:

SAS.Shared.AddIns.ISASProject
Provides access to basic project file information such as the name and the date it was created.

SAS.Shared.AddIns.ISASProcessFlow
Provides access to the contents of a process flow.

SAS.Shared.AddIns.ISASProjectTask
Represents an individual task in a process flow.

The task example shows you how to use these interfaces to navigate your project contents and to access information about the individual tasks in the project.

Example Source Files and Information

The custom task example is named SASPress.CustomTasks.TaskPropertyViewer.dll (C# version) and SASPress.CustomTasks.TaskPropertyViewerVB.dll (Visual Basic .NET version). They are both built and ready to use. Here are some details about the task examples:

.NET language and version	C# and Visual Basic, .NET 3.5 (Microsoft Visual Studio 2010)
.NET difficulty	Medium
.NET features	Windows Forms
SAS difficulty	Low
SAS features	None
Binaries	SASPress.Examples.TaskPropertyViewer.zip SASPress.Examples.TaskPropertyViewerVB.zip
Source code	SASPress.Examples.TaskPropertyViewer_src.zip SASPress.Examples.TaskPropertyViewerVB_src.zip

What's in Your Project?

The custom task example enables you to explore the contents of an active project and to peek inside some of the tasks. A useful piece of information that you might find of interest is the XML representation of the task settings (called XmlState).

In the development of a custom task, XmlState is considered the task memory. XmlState stores the settings that the user selected so that those settings are used when the task is run (for example, when the process flow is run). In addition, XmlState provides those settings to the user interface every time you open the task to modify it.

Unfortunately, the XmlState property is locked up in the active project files. It's difficult to see inside XmlState unless you attach a program debugger to the process flow or you build a tool that writes the XML content to a debug-level log file. (To learn more about debugging techniques, see Chapter 8, "Debugging Techniques: Yes, You Will Need Them.")

With the SAS Task Property Viewer, you can see the contents of the internal state as the task is saved in the project. Verifying the internal state of the task can be very useful when you are debugging a problem. Figure 10.1 shows an example of the saved task state of another custom task.

Figure 10.1: Example of SAS Task Property Viewer Showing a Saved Task Internal State

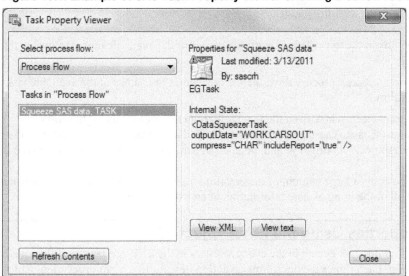

Displaying Properties in a Simple User Interface

The structure of the SAS Task Property Viewer is simple. The task class itself has almost no code in it. It contains only a few lines with the task metadata (just enough for it to show up in the application's menu). It has a simple `Show()` implementation that displays the task dialog box. Here is the C# implementation of the `Show()` method for this task, which you can find in SasTaskPropertyViewerTask.cs:

```
public override ShowResult
  Show(System.Windows.Forms.IWin32Window Owner)
{
  // this creates the dialog box to show the project/task properties
  SasTaskPropertyViewer dlg = new
    SasTaskPropertyViewer(this.Consumer);
  // NOTE: the Show() call (instead of ShowDialog())
```

```
            // opens the task dialog box in a MODELESS state, which allows you
            // to continue to interact with the project while this dialog box
            // is showing.
            dlg.Show(Owner);
            return ShowResult.Canceled;
        }
```

The class that implements the dialog box is named SasTaskPropertyViewer. This class is in the class in which the real work takes place. The user interface in the dialog box enables you to select a process flow from the active project. (Remember, a project can have multiple process flows.)

Here is how the user interface is structured:

- The combo box in the upper left displays the names of the available process flows in the project.
- When the user selects a process flow, the list box shows the names of the available tasks in the process flow.
- When the user selects a task from the list, the **Properties** area on the right shows the properties for the task. These properties include the date the task was last modified, the user ID of the person who modified it, and the internal state (XmlState) that is saved in the project.

If the internal state contains a large amount of information, the user can click one of the **View** buttons to launch a view of all of the internal state information in an external viewer.

Accessing Properties Using the ISASProject APIs

The interesting work in this task occurs in the event handlers in the user interface workflow described in the previous section. Each time a user performs an action, it leads to a bit of processing, which results in retrieving information about the project to display in the dialog box.

The main events that trigger an action are:

Refresh process flows for project
   Triggered when the task dialog box is first displayed or when the user clicks **Refresh**. The top-level list of process flows is retrieved for the active project, and the list is bound to the combo box.

Select process flow
   Triggered when the user selects a process flow from the combo box. The list of tasks in the selected process flow is retrieved and displayed in the list box.

Select task
   Triggered when the user selects a task from the list. The properties for the selected task are retrieved and displayed in the **Properties** area.

Retrieving the Process Flows

Retrieving the list of available process flows is a simple operation. With a handle to the ISASTaskConsumer3 interface, you can call GetProject() to retrieve a handle to the current project as ISASProject. With the ISASProject interface, you can retrieve a list of available process flows as ISASProjectProcessFlow objects in a generic List container.

Here is an example in C#:

```
// Consumer is the ISASTaskConsumer3 handle
// get a handle to the project, and
// then the list of process flows
ISASProject project = Consumer.GetProject();

// get the list of process flows from the project
List<ISASProjectProcessFlow> flows = project.ProcessFlows;
```

In the user interface for this task, you can use data binding to bind the list of available process flows (as ISASProjectProcessFlow objects) to the combo box. Here is a complete example of the **RefreshProjectContent()** handler in C# that retrieves the process flow list and binds it to the combo box (which is named **cmbProcessFlows**):

```
private void RefreshProjectContent()
{
    // get a handle to the project, and
    // then the list of process flows
    ISASProject project = Consumer.GetProject();

    // get the list of process flows from the project
    // and bind the list as a data source for the
    // combo box
    List<ISASProjectProcessFlow> flows = project.ProcessFlows;
    cmbProcessFlows.DisplayMember = "Name";
    cmbProcessFlows.DataSource = flows;
    cmbProcessFlows.SelectedIndex = 0;
}
```

Using data binding, the list of available process flows is automatically bound to the combo box. Whether the project contains one process flow or 20, they are all available in the combo box. Setting the **DisplayMember** property to **Name** tells the combo box to use the ISASProjectProcessFlow.Name property to populate the display text in the combo box.

Retrieving the Tasks in a Process Flow

When the user selects a process flow from the combo box, it's time to update the list of tasks with tasks that are available in the process flow.

Because you used data binding to populate the process flows in the combo box, each item in the combo box is a full-fledged .NET object that represents the ISASProjectProcessFlow object. As a result, retrieving the list of tasks in a process flow is easy. Here is a C# example:

```
ISASProjectProcessFlow flow =
   cmbProcessFlows.SelectedItem as ISASProjectProcessFlow;
List<ISASProjectTask> tasks = flow.GetTasks();
```

The next step is to populate the list box with the tasks. Once again, you use data binding to set a list of tasks as the data source. In addition, there are a few other chores:

- Change the text of the label before the list box to be the name of the selected process flow.
- Because the list of tasks in the process flow might include *this actual task* (that is, the task that is currently active), you need to remove that task from the list. There is no need to add it to the list, and its presence might confuse the user. To remove it, use the ISASTaskConsumer3 method, **GetThisTask()**, to help facilitate the comparison and to filter out the task from the inside.

Here is a C# example of the **OnSelectedFlowChanged** handler, which retrieves the list of tasks in the selected process flow and binds the list to the list box (named **lbTasks**):

```
private void OnSelectedFlowChanged(object sender, EventArgs e)
{
    // get the current selected process flow
    ISASProjectProcessFlow flow =
     cmbProcessFlows.SelectedItem as ISASProjectProcessFlow;

    // update the label, and bind the ListBox data source
    // to the list of tasks
    lbTasks.Text =
       string.Format("Tasks in \"{0}\"", flow.Name);
    lbTasks.DisplayMember = "Name";

    // get this list of tasks in the flow, and remove
    // THIS task if it's in the list
    List<ISASProjectTask> tasks = flow.GetTasks();
    if (tasks.Contains(Consumer.GetThisTask()))
        tasks.Remove(Consumer.GetThisTask());
    lbTasks.DataSource = tasks;
}
```

Selecting a Task

Until now, the event handlers have been all about navigation. They have listed the available process flows and tasks so that you can select the task that you want to peek inside of. Once you've selected a task by clicking on a task name in the list box, it's time to update the properties in the **Properties** area.

Retrieving the Properties for the Selected Task

Data binding in the list box helps make this process easier. The selected task can be cast or treated as an ISASProjectTask object. However, the ISASProjectTask interface provides access to task-specific properties in the project. There is a more generic interface (ISASProjectItem) that provides access to properties that are common to all elements in the project (including tasks, data, programs, and more). These properties include the name of the item, the date it was last modified, the user ID of the person who modified it, and the icon (image) that is used to represent it in the project. Any object that supports ISASProjectTask also supports ISASProjectItem, but you need to reference its ISASProjectItem face to access these common properties.

Here is a C# example that shows how to access the task-specific properties as well as the common properties. When you access the common properties, the code casts the selected object as a ISASProjectItem object.

```
ISASProjectTask task = lbTasks.SelectedItem as ISASProjectTask;
ISASProjectItem item = lbTasks.SelectedItem as ISASProjectItem;
if (task != null && item!=null)
{
    // hide the No Task panel
    panelNoTask.Visible = false;
    panelNoTask.SendToBack();

    // set up the values for the properties
    gbProperties.Text = string.Format("Properties for \"{0}\"",
        item.Name);
    pbTaskIcon.Image = item.GetIcon(false).ToBitmap();
    lblLastModified.Text = string.Format("Last modified: {0}",
        item.DateModified.ToShortDateString());
    lblModifiedBy.Text = string.Format("By: {0}",
        item.ModifiedByUserId);
    lblTaskType.Text = task.TaskType;
    txtInternal.Text = task.InternalState;
    btnViewText.Enabled =
       !string.IsNullOrEmpty(task.InternalState);
    btnViewXml.Enabled =
       !string.IsNullOrEmpty(task.InternalState);
}
else
{
    gbProperties.Text = "No selected task";
    panelNoTask.Visible = true;
    panelNoTask.BringToFront();
}
```

More Possibilities with SAS Enterprise Guide Projects

The task example in this chapter is simple, but useful, for developers as they test their custom tasks and diagnose problems. However, the series of ISASProject interfaces enable you to do much more to programmatically discover and influence what's in your project.

The ISASProjectProcessFlow interface is especially powerful because it enables you to test the relationships between different project items and even make changes. The following list describes methods that you can use to discover and change the user-defined links between the project items in a process flow:

AddLink
Adds a process flow link to connect two items in the project.

AreLinked
Checks to see whether two project items are directly connected with a user-specified link. The order of the parameters is unimportant—the method checks for links going in either direction.

CanLink
Checks to see whether it is possible to add a link between two project items.

SAS Enterprise Guide does not allow circular links, so this method will return a false value if the proposed link is a circular link. In addition, it returns a false value if the items are already directly linked or if one of the items is a type that does not allow links (for example, a SAS log item).

IsAncestor
Checks to see whether one ISASProjectItem is an ancestor to another ISASProjectItem in a process flow.

RemoveLink
Removes a process flow link that connects two items in the project.

In addition, there are methods that you can use to retrieve the SAS program code that is represented in your process flow:

GetAllSasCode
Retrieves all of the SAS program code that is in (or is generated by the tasks in) the current process flow.

GetSasCodeForThisPath
Retrieves all of the SAS program code that is in (or is generated by the tasks in) the current process flow. This includes code up to a given task that is passed in. It **does not** include sibling tasks or ancestors of sibling tasks that are not also ancestors of the given task. In other words, this retrieves SAS program code for a single branch of the process flow.

GetSasCodeUpToHere
Retrieves all of the SAS program code that is in (or is generated by the tasks in) the current process flow. This includes code up to a given task that you specify.

Chapter Summary

The ISASProject interfaces provide access to the internals of your SAS Enterprise Guide projects. You can use these interfaces to see inside other tasks that are in the project, navigate the project structure, and retrieve valuable information about how the project is put together and the SAS programs that it yields.

Chapter 11: Calculating Running Totals

About This Example 142
    Example Source Files and Information 142
Designing the Task Features 142
    Assumptions: They Are Necessary 143
    Scenario 1: Calculate the Running Total for One Measure across All Rows ... 143
    Scenario 2: Calculate Running Totals across Groups 144
Designing the User Interface 146
    Assembling the User Interface 147
    Hooking the Controls to Data and Events 151
Saving User Selections 157
    Using LINQ to Create XML 157
    Using LINQ to Read XML 159
Generating a Correct SAS Program 160
    Creating a Readable Program Header 160
    Applying the Task-Specific Filter 161
    Wrap Your Variable Names Appropriately 162
Chapter Summary 163

When working in SAS, there are certain programming operations that seem to be needed again and again. The problem is that some of these operations are needed only once in a while, and every time the operation is needed, you have to go back to the documentation, examples, and conference papers to figure out how to code it. A custom task can capture the recipe for how to code the operation and get the results, once and for all, and can save you valuable research time.

The task presented in this chapter performs a basic data transformation to calculate running totals of a measure in your data. It's an operation that is uniquely suitable for the SAS programming language. It's a common need, but whenever I'm faced with it, I always have to relearn how to code it.

142 Custom Tasks for SAS Enterprise Guide Using Microsoft .NET

If you are a SAS programmer, the SAS code to perform this operation is trivial. But, non-programmers might have a difficult time knowing where to begin. With a custom task, you can provide a user interface that makes the process easy for everyone.

About This Example

The task example adapts a simple SAS program and places an intuitive user interface on top of it. The user interface incorporates familiar elements from the tasks that are built in SAS Enterprise Guide, including the variable selector and the output data selector. These familiar elements help ensure that the users of the task will have no trouble learning how to use it.

This chapter describes most of the main steps to build this task, but it does not provide a complete step-by-step tutorial. Use the provided source code in this chapter as a reference, and use the content in this chapter to provide additional context and an explanation for how the task is put together.

Example Source Files and Information

The task example is named SASPress.Examples.RunningTotals.dll and is written in C#. It is built and ready to use. Here are some details about the task example:

.NET language and version	C# and Visual Basic, .NET 3.5 (Microsoft Visual Studio 2010)
.NET difficulty	Medium
.NET features	Windows Forms, SAS variable selector, output data selector
SAS difficulty	Low
SAS features	DATA step
Binaries	SASPress.Examples.RunningTotals.zip
Source code	SASPress.Examples.RunningTotals_src.zip

Designing the Task Features

When a colleague asks you to write a program to calculate running totals, is that enough information to code the operation and produce what he's asking for? Unless you're a mind reader, the answer is probably "No." You need to fine-tune the definition of the operation before you can produce a meaningful result.

That's what you're going to do here—narrow down the operation of calculating running totals to just a few scenarios that a task can handle efficiently across a wide variety of data sources.

Assumptions: They Are Necessary

You know what they say happens when you assume that something is true? (If you don't know, you'll have to look it up as I cannot spell it out here. This is a family-friendly technical book.)

To limit the problem domain of the task, you must make some assumptions about the data. For example, when calculating running totals, you must assume that the data is already ordered in some sort of logical sequence—that is, a sequence in which it makes sense to produce a cumulative value of the measure, one record after another. (Often, the order reflects a chronological sequence based on a date or time variable, but not always.)

If you didn't make this assumption, then the task might need to also sort the data. Of course, you *could* do that part as well, but you don't want to sort the data if it's already sorted—that's a waste of system resources. Instead, you can rely on the user to make sure the data records are already sorted.

SAS Enterprise Guide offers simple-to-use methods to sort data, so assume that the user has done that for you.

Scenario 1: Calculate the Running Total for One Measure across All Rows

This is the easiest scenario—it requires that you ask only one question, "Which variable do you want to use as the measure?"

For example, suppose you have a data set that contains the sales transactions for beer sales over a period of time. You want to see the cumulative total of sales over the time period that the data represents. All you really need to know about the data is this, "Which column contains the amounts of the transactions?" In this example, that column is named Sales. Here's the simple DATA step code to produce the running total:

```
data work.beertotals;
  set beer.beer_sales;
  total_sales + sales;
run;
```

Figure 11.1 shows a SAS Enterprise Guide session with the program and the output data set, complete with the new column.

Figure 11.1: Calculating the Running Total Using SAS DATA Step Code

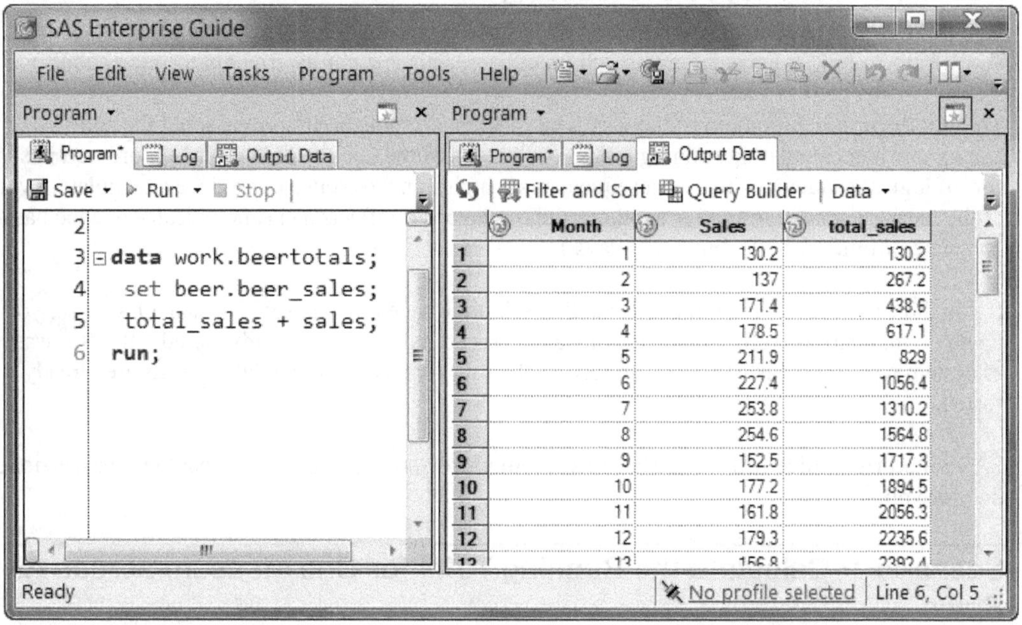

```
data work.beertotals;
  set beer.beer_sales;
  total_sales + sales;
run;
```

Month	Sales	total_sales
1	130.2	130.2
2	137	267.2
3	171.4	438.6
4	178.5	617.1
5	211.9	829
6	227.4	1056.4
7	253.8	1310.2
8	254.6	1564.8
9	152.5	1717.3
10	177.2	1894.5
11	161.8	2056.3
12	179.3	2235.6
13	156.8	2392.4

Scenario 2: Calculate Running Totals across Groups

The other common scenario for running totals occurs when your data are segmented across a grouping variable. The grouping variable might be a time dimension—you have daily transactions that you want to sum to the month level. Or, your data might be subject-related—you have clinical trial data that's grouped by patient ID, and you want to measure the cumulative effect of a treatment.

To accomplish these operations, you need to keep a running total for the measure that you want. And, you need to reset that total to zero each time you encounter a new value for the grouping variable. The SAS DATA step makes this very easy by creating an automatic variable named FIRST.*variable*, where *variable* is the name of the grouping variable. FIRST.*variable* is created only when you specify the grouping variable in a BY statement. And, as stated previously, the data needs to be sorted before you can use the grouping variable.

For example, this SAS program creates a running total of beer sales by customer ID:

```
data beer_cust_totals;
  set beer_byCust;
  by custID;
  if first.custID then
        total_sales=sales;
  else
        total_sales+sales;
run;
```

Figure 11.2 shows the program and the resulting data in SAS Enterprise Guide. When the `custID` column changes values at row 36, the `total_sales` column value resets to the baseline value from the `sales` column.

Figure 11.2: The Grouped Data with Running Totals

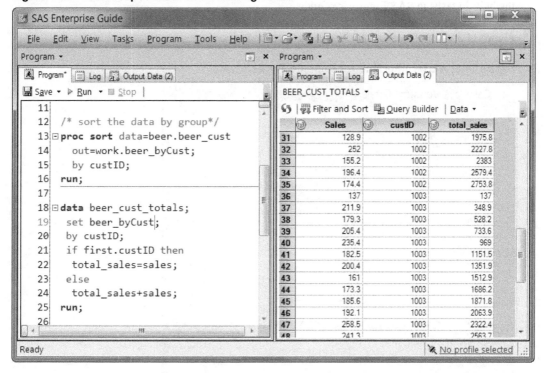

Designing the User Interface

Now that you know how to code the operation in SAS, you have enough information to design a user interface. The user interface for this task has to enable the user to specify the following information:

- Which column is the source of the running total?
- Which column (if any) serves as the grouping variable?
- What should the column containing the totals be named in the output?
- What is the name and location of the output data set to be created?

You could create a simple user interface for the user to select the responses for the first two questions, just as you did for the Top N Report in Chapter 9, "The Top N Report." However, in this example, you use the SAS variable selector control, which helps provide a look and feel similar to the built-in tasks in SAS Enterprise Guide.

Figure 11.3 shows the user interface design for the task.

Figure 11.3: A User Interface for Running Totals Task

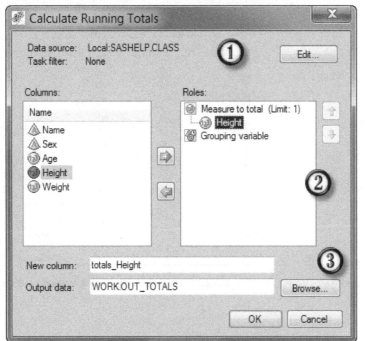

The UI design has a few key elements that make it look consistent with the user interface of many built-in tasks. These key elements are labeled in Figure 11.3.

❶ The input data source is displayed with an **Edit** button so that the user can make changes. Not only can the user select a different input data source, the user can also specify a task-specific filter that is applied when the task is run. All of these features are implemented in a common control called the TaskSelectedDataControl.

❷ The variable selector control (known as SASVariableSelector) shows the available columns in the input data and enables the user to assign variables to task-specific roles.

❸ The location of the output data set is shown in a standard text field. The **Browse** button invokes the application-specific File Save dialog box, which enables the user to navigate through potential output locations in SAS libraries.

Assembling the User Interface

This project uses the standard project structure that the selected SAS template creates for you. To begin assembling the user interface, create a new project using a Microsoft Visual Studio template as described in Chapter 3, "Creating Custom Task Projects in Microsoft Visual Studio." You can use a C# template or a Visual Basic .NET template, based on your preference. Name your project RunningTotals.

Step 1: Add the Necessary Assembly References

Before you add any substance to your project (beyond the boilerplate classes that the template creates for you), you need to reference two additional assemblies (DLL files) that contain the UI components that you need:

- SAS.SharedUI.dll
- SAS.EG.Controls.dll

You learned how to add references to DLL files in earlier chapters. But, here's a quick reminder of the process—right-click on the project in the Solution Explorer view, and select **Add Reference**. On the left, click the **Browse** tab, navigate to the folder in which SAS Enterprise Guide is installed, and select these two DLL files.

You will be using the classes only in the SAS.EG.Controls.dll file, but those classes have dependencies on other classes in other assemblies, including classes in the SAS.SharedUI.dll file. For your project to build successfully, you need to reference this additional assembly.

> SAS documents many of the classes that are included in the SAS.EG.Controls.dll file. However, not every class is documented and supported for you to use. If you use an undocumented class, you might find that your task doesn't work correctly with a future release of SAS Enterprise Guide.

In the Solution Explorer view, select both DLL files, and change **Copy Local** to **False** in the Properties view. Changing this value prevents the build process from copying several DLL files into your output data set folders.

Step 2: Clean Up the Boilerplate User Interface

The Microsoft Visual Studio template creates a few placeholder controls that you don't need in this project, so you can delete them.

1. Open RunningTotalsTaskForm.cs in the Design view. The task form displays default TextBox and Label controls provided by the template.
2. Select the **TextBox** and **Label** controls, and delete them. You'll be adding different controls for this task.
3. Delete the references to the **TextBox** and **Label** controls from RunningTotalsTaskForm.cs. Open the RunningTotalsTaskForm.cs program, and delete the line that references the **textBox** control and the line that references the **label** control.
4. Your project should build cleanly at this point. Verify that it builds cleanly by selecting **Build→Rebuild Solution**.

Step 3: Add Common Task Controls to the Toolbox

To use the built-in controls in the Visual Studio Form Designer, you must add those controls to the Toolbox view. This enables you to add SAS controls to a form just as you add Windows controls such as a textbox or button.

1. Open RunningTotalsTaskForm.cs in the Design view. If the Toolbox view is not visible, select **View→Toolbox**.
2. Right-click in the Toolbox view, and select **Add Tab**. This creates a new section in the **Toolbox** list, and that's where you'll group the SAS controls. Enter SAS as the name of the new tab.
3. With the mouse pointer under the new **SAS** tab, right-click, and select **Choose Items**. The Choose Toolbox Items dialog box appears.
4. With the **.NET Framework Components** tab selected, click **Browse**. The Open window appears.
5. Navigate to the folder in which SAS Enterprise Guide is installed, and select SAS.Tasks.Toolkit.dll. Click **Open**. The list of components now contains all of the components that are provided in the SAS Task Toolkit.
6. To see the list of SAS components grouped together, click the **Assembly Name** column heading, and then scroll to the **SAS.Tasks.Toolkit** grouping. Figure 11.4 shows an example of this dialog box.
7. All of the available controls are selected (as indicated by the check boxes), but, for this task, you need only the **TaskSelectedDataControl**. You can either keep all of the controls that are selected, or you can uncheck all of them except **TaskSelectedDataControl**. Either way works.
8. Click **OK** to close the Choose Toolbox Items dialog box. The new controls are now shown in the **SAS** grouping in the Toolbox view.

Figure 11.4: The Choose Toolbox Items Dialog Box with New SAS Controls

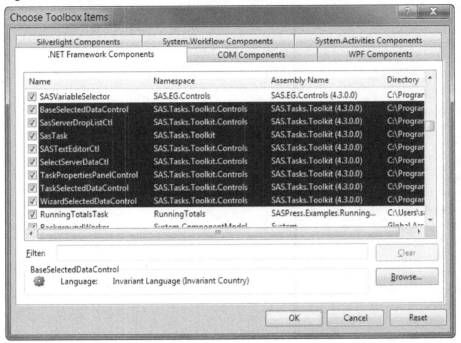

Repeat the previous steps, but, this time, select SAS.EG.Controls.dll.

There are many controls in SAS.EG.Controls.dll. In a future task, you might want to use them. For now, uncheck all of them except **SASVariableSelector**, and then click **OK**.

After you have completed both sets of steps successfully, the Toolbox view should include a list of controls similar to the ones shown in Figure 11.5.

Figure 11.5: The SAS Group of Controls in the Toolbox

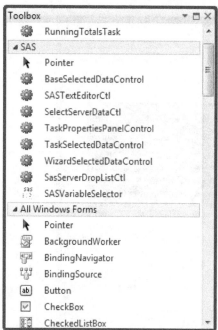

Step 4: Add the SAS Task Toolkit Controls to the Form

With the SAS controls in the Toolbox view, it's easy to add them to the task form. You can drag and drop the controls onto the form just as you can drag and drop any built-in Windows control.

The SAS controls provide a designer preview on the form. The designer preview shows you what the controls will look like when the form is live at run time, even though the controls aren't displaying any data. Figure 11.6 shows an example of what the controls look like after you place them on the form.

Figure 11.6: SAS Controls in the Design View

Hooking the Controls to Data and Events

The built-in controls help ensure a consistent user experience. They save you the trouble of having to re-implement common features in your custom task. But, they are far from automatic, as you will soon learn. Some extra programming is required to make the controls work properly for your task.

The following list includes brief descriptions of the additional .NET programming that is required to connect the dots between the controls and the logic of your task:

- Connect the active data source to the TaskSelectedDataControl, including any selected filter. When the user changes the data source or the filter, your task must react to the change.
- Define the data roles for the SASVariableSelector, and add the columns from the active data into it. If columns have already been selected and saved in the settings, you must add those columns to SASVariableSelector.
- Display the Save dialog box when the user clicks the **Browse** button to select the output data set.

Initializing the TaskSelectedDataControl

You want the name of the active data source to be displayed in TaskSelectedDataControl. The application (in this case, SAS Enterprise Guide) keeps track of the active data source, so initializing the control is simple. Here's an example in C#:

```
// initialize the data selection control with a
// handle to the consumer application
// and the active data
selDataCtl.Consumer = Consumer;
selDataCtl.TaskData = Consumer.InputData[0] as ISASTaskData2;
```

However, this initializes only part of the control. It doesn't initialize the task-specific filter settings if any have been specified. To accomplish that, you need an additional line:

```
selDataCtl.FilterSettings = Settings.FilterSettings;
```

As the task developer, it is your job to collect, save, apply, and then reset the task-specific filter if one is used. The SAS Task Toolkit provides helper methods that help you do this as you will see in later steps.

Because the TaskSelectedDataControl is somewhat a black box (because you don't have direct access to the UI components in it), you have to rely on public properties and events to communicate with it. One event that you will probably want to handle is DataSelectionChanged, which tells you that the user has changed the input data source or the filter on the data.

You can use the Design View in Visual Studio to add the event handler to your code, or you can manually add it yourself :

```
this.selDataCtl.DataSelectionChanged +=
    new System.EventHandler(this.OnDataSelectionChanged);
```

If you write your own code to subscribe to the event, place it in an initialization routine, such as InitializeComponent() or in an override implementation of the OnLoad() method.

In the event handler, you want to save the new filter settings if they were changed:

```
private void OnDataSelectionChanged(object sender, EventArgs e)
{
    FilterSettings fs = selDataCtl.FilterSettings;
    Settings.FilterSettings = fs;
}
```

Defining the Role Categories in SASVariableSelector

The role categories on the right side of the SASVariableSelector are more than just names and pretty icons. Each role category contains rules for what type of variable and how many can be assigned to the role.

You define the rules by creating a specific .NET data structure that contains the constraints. You add that data structure to the role. The following example defines the role for the measure variable in the Running Totals task. This variable is required and can accept only one value.

```
// this one is the "Measure" role, which requires a numeric variable
SASVariableSelector.AddRoleParams parms =
    new SASVariableSelector.AddRoleParams();
parms.Name = Translate.MeasureValueRole;
parms.MinNumVars = 1; // Min of 1 says required
parms.MaxNumVars = 1; // one and only one
// not going to allow "macro"-style prompts
parms.AcceptsMacroVars = false;
parms.Type = SASVariableSelector.ROLETYPE.Numeric;
varSelCtl.AddRole(parms);
```

> In this example, note the use of **Translate.MeasureValueRole**, which refers to a string resource that is stored in the .NET string resource table. Instead of hardcoding a string literal in the .NET code, this string resource provides a level of indirection to look up the value of the string from a central location.
>
> This serves two purposes. First, it ensures that the same exact value is used in all parts of your code logic. If every reference to the value is resolved using the string resource table, then you will always have the same exact value. This is important because there might be additional uses of SASVariableSelector that need to reference this value.
>
> Second, it ensures that your task is localizable, meaning that it can be translated for use in another language. All SAS tasks are localizable, and SAS Enterprise Guide is translated into at least 15 different languages.

In contrast, the grouping variable is not required, and the task can actually handle multiple groups. Here is the code to set the properties for this role:

```
// and this one is the "Grouping" role, optional
SASVariableSelector.AddRoleParams labelParm =
    new SASVariableSelector.AddRoleParams();
labelParm.Name = Translate.GroupingRole;
labelParm.MinNumVars = 0; // optional
// No labelParm.MaxNumVars to allow as many as you want
// not going to allow "macro"-style prompts
labelParm.AcceptsMacroVars = false;
labelParm.Type = SASVariableSelector.ROLETYPE.All;
varSelCtl.AddRole(labelParm);
```

Adding the Data Columns to the Variable Selector

The left side of the SASVariableSelector shows the available columns in the active data source. For columns to appear in the SASVariableSelector, you must write code that tells the SASVariableSelector about them. This can be cumbersome, so the Running Totals task example includes a helper method, named BuildVariableParamsList, that makes it easier.

The BuildVariableParamsList method maps the contents of a SAS.Task.Toolkit.SasData object to the variable list that the SASVariableSelector requires. Here's how you use it:

```
public static List<SASVariableSelector.AddVariableParams>
    BuildVariableParamsList(SasData data)
{
    // Allocate the list
    List<SASVariableSelector.AddVariableParams> parmList =
        new List<SASVariableSelector.AddVariableParams>();

    foreach (SasColumn col in data.GetColumns())
    {
        SASVariableSelector.AddVariableParams parms =
            new SASVariableSelector.AddVariableParams();

        // populate the column properties
        // from the SasColumn entry
        parms.Name = col.Name;
        parms.Label = col.Label;
        parms.Format = col.Format;
        parms.Informat = col.Informat;

        // map the column category
        // to the variable selector
        // version of this enumeration
        // Ensures the correct "type" icon is
        // shown in the variable selector
        switch (col.Category)
        {
            case SasColumn.eCategory.Character:
                parms.Type = VARTYPE.Character;
                break;
            case SasColumn.eCategory.Numeric:
                parms.Type = VARTYPE.Numeric;
                break;
            case SasColumn.eCategory.Currency:
                parms.Type = VARTYPE.Currency;
                break;
            case SasColumn.eCategory.Date:
                parms.Type = VARTYPE.Date;
                break;
            case SasColumn.eCategory.DateTime:
                parms.Type = VARTYPE.Time;
                break;
            case SasColumn.eCategory.Georef:
                parms.Type = VARTYPE.GeoRef;
                break;
```

```
            default:
                parms.Type = VARTYPE.Numeric;
                break;
        }
        // add the complete object to the list
        // to return
        parmList.Add(parms);
    }
    return parmList;
}
```

The BuildVariableParamsList method can be used in more generic ways in any custom task where you use the SASVariableSelector. In the following example, the method is included in a static class named **Helpers** (in the Helpers.cs file). Here's how you use it in C#:

```
SasData data = new SasData(
    this.Consumer.InputData[0] as ISASTaskData2
    );

// this builds the variable list
 // in the format that the variable selector needs
 List<SASVariableSelector.AddVariableParams> parmList =
     Helpers.BuildVariableParamsList(data);
 varSelCtl.AddVariables(parmList);
```

Remembering Where You Parked Your Variables

If your user is modifying an existing instance of this task, it's not enough to add the variables from the data set. You also have to add the variable assignments. To accomplish this, you must read the saved variable assignments from the task settings, and then use code to reassign them.

This example has only two roles—one role that is required (for the measure), and one role that is optional (for grouping). The following C# code is used to reassign the variables after they are read into your task's Settings object:

```
// assign the selected measure, if any
 if (!string.IsNullOrEmpty(Settings.VariableMeasure))
 {
     if (data.ContainsColumn(Settings.VariableMeasure))
         varSelCtl.AssignVariable(Translate.MeasureValueRole,
             Settings.VariableMeasure);
 }

// assign the selected group vars, if any
 foreach (string var in Settings.VariableGroups)
 {
     if (data.ContainsColumn(var))
         varSelCtl.AssignVariable(Translate.GroupingRole, var);
 }
```

Notice how the code doesn't take much for granted. It checks to make sure that the saved variable name still exists in the input data (using the `data.ContainsColumn` method) because a data set *can change* between uses. To be complete, this code should be wrapped in a Try and Catch block to catch SASVariableSelector.VariableNotAssignedToRole exceptions (in case something else goes awry). What happens if a measure variable was assigned that turns out to be a character variable? Anything can happen in the world of SAS, and it can take a lot of work to be ready for it.

Validating the User Selections

What is the best way to handle a data entry error when your task runs? Don't allow the data entry error in the first place! By putting simple validation checks in place, you can prevent the user from specifying an invalid name for the output data set or the column for totals.

Regular expressions provide a convenient (although cryptic) method for ensuring that a text value follows the SAS syntax rules. In .NET programming, the regular expression capability is in the System.Text.RegularExpressions.Regex class. Here is an example that checks whether the new column for totals has a valid SAS name:

```
private bool isValidVarName(string name)
{
    // a regular expression to match a variable name
    // 1-32 chars, begin with alpha, then alphanumeric
    // or underscores
    Regex regex = new Regex(
        "/^[a-z_]\w{0,31}$/i",
        RegexOptions.Compiled);
    return (regex.IsMatch(txtTotalsCol.Text));
}
```

The details about how regular expressions work could fill an entire book. Briefly, the expression in this example ensures that the text value follows these SAS syntax rules:

- Has between 1 and 32 characters.
- Starts with an uppercase letter, a lowercase letter, or an underscore.
- Continues with zero or more (up to 31) alphanumeric characters or underscores.

In your task form, you should check text values before allowing the user to click **OK**. This ensures that at least at the time the task form was closed, all of the text fields in the form contained valid text values. In the source code for the Running Totals task example, the **OK** button is disabled until all controls and fields contain valid data.

Saving User Selections

As you have seen in the other chapters, storing user selections is an important part of creating a useful task. The task life cycle (described in Chapter 4, "Meet the Custom Task APIs") relies on these saved settings when it comes time to generate a SAS program and repopulate the user interface.

Like many of the other examples, this task example relies on a settings class in .NET programming to save the values of the user selections in an XML format.

The task in this chapter is a simple task with just a few user selections to remember:

- the measure variable
- the grouping variables
- the output data set name
- the name of the column for totals

Because this task supports a task-specific filter, you also need to remember the filter settings.

Using LINQ to Create XML

In this task example, you use the LINQ (language-integrated query) feature of the .NET Framework to build and parse the XML. LINQ adds native data querying capabilities to everyday .NET data structures, including the XML document object.

In the following examples, you use the LINQ classes XDocument (to wrap the XML structure) and XElement (to store each value from the Settings object). Here is the C# implementation of the `ToXml()` function to show how these classes can be used. This example can be found in the RunningTotalsSettings.cs file.

```
using System.Xml.Linq;

int version = 1;
public string ToXml()
{
 // using LINQ to save to Xml
 // create an "inner" element to hold
 // the list of grouping variables
 // could be 0 or more
 XElement groupsElement = new XElement("GroupVariables");
 foreach (string var in VariableGroups)
 {
    groupsElement.Add(new XElement("GroupVariable",var));
 }
```

```
    XDocument doc = new XDocument(
       new XDeclaration("1.0", "utf-8", string.Empty),
       new XElement("RunningTotalsTask",
          new XElement("Version", version),
          new XElement("FilterSettings", FilterSettings.ToXml()),
          new XElement("DataOut", DataOut),
          new XElement("MeasureVar", VariableMeasure),
          new XElement("TotalsVar", VariableTotal),
               groupsElement
       )
       );

    return doc.ToString();
}
```

The LINQ statements to construct the XML are easy to read. You can almost guess what the resulting XML structure will look like. Each XElement object represents an XML tag with the value that it stores. Nested XElement objects represent the hierarchical structure of the XML that you are storing.

> Did you notice the **version** element in the XML structure? It's important to keep up with the version of your task XML to help with task maintenance in the future. Regardless of how satisfied your users are with your task, they always ask for enhancements. If you decide to support new elements in a new version of your task, the task logic will need to determine whether the task XML contains these new elements or whether the task XML was created by a previous version of your task. By adding a **version** element, you create a reliable way to determine which values you can glean from the XML.

Remember the SAS Task Property Viewer from Chapter 10, "For the Workbench: A SAS Task Property Viewer?" You can use the SAS Task Property Viewer here to verify that the XML is formatted with the fields that you expect. Figure 11.7 shows an example of the task XML that is created by the previous LINQ statements.

Figure 11.7: Example of the Running Totals Task XML in the SAS Task Property Viewer

Using LINQ to Read XML

To read the task XML back into your Settings object, you, again, use the XDocument and XElement classes to parse the XML tags back into the intrinsic types of values that you need. The XDocument class has a Parse method that loads the XML into the object. You can then use the XElement class to capture each element's setting by its tag name. Here is the C# implementation of the **FromXml()** function to read the XML and populate the values of the Settings object. This example can be found in the RunningTotalsSettings.cs file.

```
public void FromXml(string xml)
{
 try
 {
  // and use LINQ to parse it back out
  XDocument doc = XDocument.Parse(xml);

  XElement filter = doc
    .Element("RunningTotalsTask")
    .Element("FilterSettings");
  FilterSettings = new FilterSettings(filter.Value);

  XElement outdata = doc
    .Element("RunningTotalsTask")
    .Element("DataOut");
  DataOut = outdata.Value;

  XElement measure = doc
    .Element("RunningTotalsTask")
    .Element("MeasureVar");
```

```
      VariableMeasure = measure.Value;

      XElement totals = doc
        .Element("RunningTotalsTask")
        .Element("TotalsVar");

      VariableTotal = totals.Value;

      XElement groups = doc
        .Element("RunningTotalsTask")
        .Element("GroupVariables");

      var g = groups.Elements("GroupVariable");
      foreach (XElement e in g)
      {
       VariableGroups.Add(e.Value);
      }

    }
    catch (XmlException)
    {
     // couldn't read the XML content
    }
  }
```

Generating a Correct SAS Program

When the scope of this task was originally defined, there were two main scenarios that it was designed to support: calculate a running total value for a single column, and calculate a running total value for a column where you can reset the running total value for each new grouping.

In either scenario, the resulting program is going to be short—just a few lines of SAS DATA step code. Therefore, it shouldn't be too much trouble to generate the code statements inline in the task logic. (Compare this approach with the approach that you took in the Top N Report in Chapter 9, where you embedded a .SAS file as a program template in the task.)

The GetSasProgram() method in the RunningTotalsSettings.cs file contains all of the logic to build the SAS program based on the selected settings. The entire method is not included in this chapter, but a few of the key elements are.

Creating a Readable Program Header

If you look at the SAS programs that most SAS Enterprise Guide tasks create, you will probably notice a standard comment block at the top of each program. This comment block contains the name of the task, the date and time the program was generated, and a few other details. Here's an example from the Scatter Plot Matrix task:

```
/* ----------------------------------------------------------------
   Code generated by SAS Task

   Generated on: Saturday, September 08, 2012 at 11:03:53 AM
   By task: Scatter Plot Matrix

   Input data: SASApp:SASHELP.CARS
   Server: SASApp
   ---------------------------------------------------------------- */
```

The SAS Task Toolkit library provides a helper method that makes it easy to create a similar program header in your task. Here's an example of how this helper method is used in the Running Totals task:

```
// this handy method creates an
// easy-to-read comment block
// in the generated SAS program
program.Append(
    SAS.Tasks.Toolkit.Helpers.UtilityFunctions.BuildSasTaskCodeHeader(
    "Calculate Running Totals",
    string.Format("{0}.{1}",consumer.ActiveData.Library,
        consumer.ActiveData.Member),
    consumer.AssignedServer));
```

Applying the Task-Specific Filter

Earlier in this chapter when you created the task's user interface, you worked hard to include the TaskSelectedDataControl. This was so that the user could apply a task-specific filter. When you generate the SAS program, you need to make sure that the filter is applied if it was specified.

The FilterSettings class provides a method named FilterValue that produces an expression of the filter selections in SAS syntax. In this task, you include that syntax in the WHERE= data set option when you specify the SET statement for the input data. Here's an example of how to get the filter selections and build the WHERE= data set option:

```
string filter = "";
if (!string.IsNullOrEmpty(FilterSettings.FilterValue))
    filter = string.Format("(where=({0}))",
        FilterSettings.FilterValue);

// then later, applying the filter in SET
program.AppendFormat("data {0};\n", DataOut);
program.AppendFormat("   set {0}.{1}{2};\n",
    consumer.ActiveData.Library,
        consumer.ActiveData.Member,
        filter);
```

For example, if the user specifies the filter shown in Figure 11.8, the generated SAS program would look similar to this:

```
data WORK.OUT_TOTALS;
  set SASHELP.CARS(
    where=(Origin IN ('Europe','USA') AND Cylinders >= 6)
  );
  totals_MSRP + MSRP;
run;
```

Figure 11.8: Example of a Task-Specific Filter

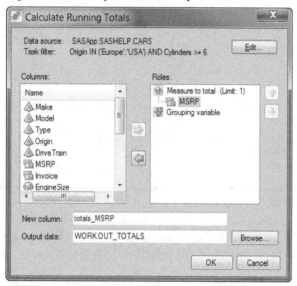

Wrap Your Variable Names Appropriately

Earlier in this chapter, you learned how to validate the name of the column for the running totals value. Despite your best efforts, not all variable names follow the old-style SAS naming rules. Not to fear, SAS has an option for that.

The option is called VALIDVARNAME=ANY. When this option is enabled, SAS allows you to use variable names that contain blank spaces or special characters. However, SAS requires that you express these variable names using special name literal syntax, which includes quoting the name and adding **n** as a name indicator. For example, you can have a variable named My Crazy Variable Name! using the following SAS code:

```
length "My Crazy Variable Name!"n 8;
```

Although most SAS programmers don't willingly use variable names like this, SAS Enterprise Guide makes it easy for you to create such names unwittingly. For example, when you import data from a Microsoft Excel spreadsheet, the column headings might not comply with SAS naming rules. Because these column headings automatically become variable names in your data set, SAS Enterprise Guide generates the proper literal syntax automatically.

As a task developer, you have to make sure that you handle these types of variable names because they might be in the task's input data. The SAS Task Toolkit library provides a helper method that automatically uses the name literal syntax. The helper method is named SAS.Tasks.Toolkit.Helpers.UtilityFunctions.SASValidLiteral(). Here's an example of how to use it:

```
using SAS.Tasks.Toolkit.Helpers;
/* ... */
program.AppendFormat("   {0} + {1}; \n",
    UtilityFunctions.SASValidLiteral(VariableTotal),
    UtilityFunctions.SASValidLiteral(VariableMeasure));
```

The **SASValidLiteral** function returns the name of the variable as is, unless the variable name doesn't comply with SAS naming rules. In that case, the function quotes the name and adds the **n** name indicator. It also handles other little evil nuances, such as the variable name *containing a quotation mark*. In that case, it escapes the quotation mark and keeps the SAS language parser happy.

Chapter Summary

The Running Totals task is an example of a classic task in SAS Enterprise Guide. It accepts input data, generates a SAS program, and produces output that can be used in a subsequent analytics or reporting step.

You learned how to incorporate many of the standard elements that SAS Enterprise Guide users expect, including the TaskSelectedDataControl, the SASVariableSelector, and the output data selector. By using these familiar elements, you make it easier for users to get started with your task. Your task has a user interface that they already know how to use.

Chapter 12: Abracadabra: Turn Your Data into a SAS Program

About This Example	**166**
Example Source Files and Information	166
Dissecting a SAS Data Set	**167**
Using .NET to Read Data from SAS Data Sets	**167**
Creating an Elegant Task Flow	**169**
Adding the SAS Enhanced Editor to a Windows Form	**170**
Using ISASTaskExecution to Take Matters into Your Own Hands	**172**
Cancel: Support Is Optional	**173**
ResultCount: How Many Results?	**174**
Run, Task, Run!	**174**
Chapter Summary	**176**

The DATA step is at the heart of the SAS programming language. The statements in a DATA step are the ingredients that, when combined and cooked by a RUN statement, create a data set that you can serve now and use for further analysis.

Have you ever wanted to take a fully cooked data set and reverse-engineer it back into its basic recipe? The task presented in this chapter enables you to do just that—you can return your SAS data set back to a SAS program.

About This Example

This task example demonstrates several techniques that you can apply in your own tasks, including:

- Use ADO.NET to connect to and read SAS data. (ADO.NET is the method for working with data sources in .NET programs.)
- Use SAS DICTIONARY tables to discover data set and column attributes for the source data. Then, use that information to influence task behavior.
- Use the SASTextEditorCtl control from SAS.Tasks.Toolkit.Controls to preview the SAS program in the SAS color-coded Program Editor.
- Encapsulate the meat of the task—the business logic that reads data and creates a SAS program—into a separate .NET class. This makes the task more maintainable and enables you to reuse the business logic in other contexts.

Use a special task interface, named ISASTaskExecution, to implement the work of the task. This enables SAS Enterprise Guide to delegate all task processing to your task so that it can perform work that can't be done easily in a SAS program.

Example Source Files and Information

The task example is named SASPress.CustomTasks.DS2Datalines.dll and is written in C#. It is built and ready to use. The source code is in SASPress.CustomTasks.DS2Datalines.sln, which is a Microsoft Visual Studio 2008 solution. Here are some details about the task example:

.NET language and version	C# and Visual Basic .NET 2.0 (Microsoft Visual Studio 2008 and later)
.NET difficulty	Medium
.NET features	Windows Forms, ADO.NET with OLE DB data providers, ISASTaskExecution processing, reuse the SAS Program Editor as a .NET control
SAS difficulty	Medium
SAS features	DATA step, SAS DICTIONARY tables
Binaries	SASPress.CustomTasks.DS2Datalines.zip
Source code	SASPress.CustomTasks.DS2Datalines_src.zip

Dissecting a SAS Data Set

To reverse-engineer a data set into a SAS program, you need to gather descriptive information about the data set itself, the columns that it contains, and the actual data records. The descriptive information includes:

- Data-set-level attributes, such as the data set label, passwords, indexes, and integrity constraints. In the task example in this chapter, passwords, indexes, and integrity constraints are skipped. The example shows how to gather and reconstruct the data set label. You can use that approach to gather information about the other attributes.
- Column attributes, such as the name, type, and length of each column, its associated SAS format and informat, and its ordinal position in the data set.

This metadata information is available in a set of SAS tables called DICTIONARY tables. And, guess what? You can read that information the same way you read a SAS data set. You can read a set of SAS tables to get information about other SAS tables.

(To learn more about DICTIONARY tables in SAS, visit support.sas.com, and search for `DICTIONARY tables`. You'll find reference documentation, several samples, and conference papers.)

The final ingredient for reconstructing a SAS data set is data values. To get the data values, open a handle to the SAS data set and read the values. SAS data values have two representations—formatted values and unformatted (or raw) values. For the purpose of building a SAS program, it doesn't matter which representation you use as long as you don't lose precision along the way. Often, a formatted value is less precise than a raw value.

Using .NET to Read Data from SAS Data Sets

Reading SAS data is one of the basic operations that custom tasks often need to perform. Most tasks in SAS are data-centric.

From .NET programs, you can open and read SAS data using a technology called ADO.NET. ADO.NET contains classes that enable you to read a variety of different data sources consistently using an intermediary piece called a *data provider*. SAS provides several data providers to connect to SAS data in different ways. In this example, you use the SAS IOM OLE DB data provider. You don't need to know much about the OLE DB protocol to use it, as you'll see.

Here are the basic steps for using ADO.NET and OLE DB to access SAS data:

1. Create a new OLE DB connection to SAS. To create a connection, you need to build a connection string—a special incantation that contains the name of the data provider and some information about what exactly you are connecting to. SAS.Tasks.Toolkit provides a simple method in the SasServer class that creates the connection string for you. For example, in C#:

```csharp
SasServer sasServer = new SasServer("SASApp");
OleDbConnection connection = sasServer.GetOleDbConnection();
```

2. Using the connection, you specify a command, which is a query that specifies which data you want to read. You can execute that command using a data reader object. For example, using an established connection, you can specify a query to capture the data set label:

```csharp
// use SASHELP.VTABLE to get data set options,
// such as data set label
string selectclause = string.Format("select memlabel
    from sashelp.vtable where
    libname='{0}' and memname='{1}'",
    _libref, _member);
OleDbCommand command =
    new OleDbCommand(selectclause, connection);
string dsLabel = "";
using (OleDbDataReader dataReader =
        command.ExecuteReader())
{
  if (dataReader.Read())
  {
    dsLabel =
      dataReader["memlabel"].ToString();
  }
}
```

The SASPress.CustomTasks.DS2Datalines.sln example includes a .NET class with the methods for reading the data set metadata and the data observations. The class is named SASPress.CustomTasks.DS2Datalines.DatasetConverter and it's stored in DatasetConverter.cs. Most of the real work that this task performs is encapsulated in this class.

The class itself is not tied to the task. You can easily use this class in another custom task project. Here is a simple C# program that uses the class to create a SAS program from SASHELP.CLASS, which is located on the SAS server named SASApp.

```csharp
DatasetConverter dc =
    new DatasetConverter("SASApp", "SASHELP", "CLASS");
string program =
    dc.GetCompleteSasProgram("WORK.CLASS");
MessageBox.Show(
    string.Format("Read {0} columns and {1} records.",
    dc.ColumnCount, dc.RowCount));
```

In addition to producing a complete SAS program, the class contains a helper method for returning just the header of the DATA step. The header is a program without all of the records embedded as data lines. The header is useful for previewing the structure of the program without reading all of its records. This task performs well even on large data. But, imagine reading all of the data values from a data set with thousands or millions of records—it's going to take some time.

Creating an Elegant Task Flow

The user interface for the task example is simple and contains just a few basic elements:

- A preview area, which uses the SAS Enhanced Editor to provide a glimpse of what the resulting program will look like.
- A text field and **Browse** button to show the output data set name and allow the user to change it.
- A **Copy to clipboard** button that lets you copy the DATA step header to the Windows clipboard.
- The usual **Run** and **Cancel** buttons.

Figure 12.1 shows the task's user interface as it's laid out in the Design view in Visual Studio.

Figure 12.1: The Task's User Interface in the Design View

Adding the SAS Enhanced Editor to a Windows Form

You can enhance your own tasks with the SAS Enhanced Editor, the fancy program editor that is provided with SAS and SAS Enterprise Guide. The SAS Enhanced Editor is a shared component that technically can be used in a variety of Windows programs. SAS does not typically document and support the use of the SAS Enhanced Editor outside of SAS applications.

SAS provides the SAS.Tasks.Toolkit.Controls.SASTextEditorCtl helper class, which hides the complexity of using the SAS Enhanced Editor. This helper class provides a few simple methods to enable you to display SAS programs and SAS log content.

To add SASTextEditorCtl to your project, you must first add the control to your Visual Studio toolbox. After the control is registered in the Toolbox view, you can add it to your Windows Form by dragging and dropping it onto the form in the Design view.

To add SASTextEditorCtl and other SAS.Tasks.Toolkit controls to your toolbox:

1. Right-click in the Toolbox view, and select **Add Tab**. This creates a new section in the **Toolbox** list, and that's where you'll group the SAS controls. Enter SAS as the name of the new tab.
2. With the mouse pointer under the new **SAS** tab, right-click, and select **Choose Items**. The Choose Toolbox Items dialog box appears.
3. With the **.NET Framework Components** tab selected, click **Browse**. The Open window appears.
4. Navigate to the folder in which SAS Enterprise Guide is installed, and select SAS.Tasks.Toolkit.dll. Click **Open**. The list of components now contains all of the components that are provided in the SAS Task Toolkit.
5. To see the list of SAS components grouped together, click the **Assembly Name** column heading, and then scroll to the **SAS.Tasks.Toolkit** grouping. Select each component that you want to make available in the toolbox. Figure 12.2 shows an example of this dialog box.

Chapter 12: Abracadabra: Turn Your Data into a SAS Program **171**

Figure 12.2: The Choose Toolbox Items Dialog Box with New SAS Controls

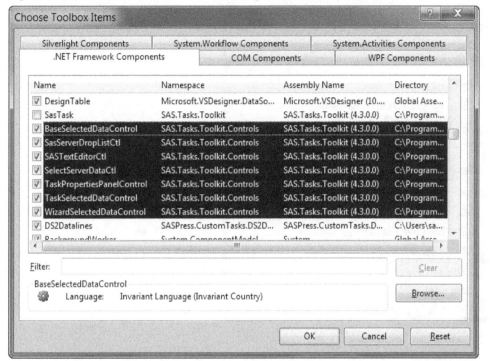

6. With the DS2DatalinesForm.cs file in the Design view, select SASTextEditorCtl, and drag and drop it onto the form. You can position and size the control just as you position and size any built-in .NET control. Figure 12.3 shows the Toolbox view with the new SAS controls.

Figure 12.3: Toolbox View with New SAS Controls

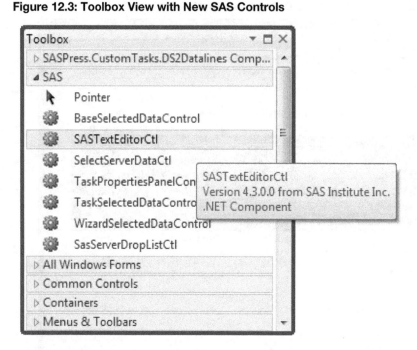

Using ISASTaskExecution to Take Matters into Your Own Hands

Unlike a traditional task, this task example doesn't generate a SAS program, and then rely on SAS Enterprise Guide to run it and generate results. Instead, the main work of this task is to read the contents of the data set and produce a SAS program that is, in itself, the result of the task.

The task needs to perform special actions when SAS Enterprise Guide gives it the cue to run. For SAS Enterprise Guide to know about this arrangement, the task class must implement a special contract called ISASTaskExecution.

The ISASTaskExecution interface includes hooks for SAS Enterprise Guide to run the task, gather customized results (a SAS program, in this case), and create a text log. Here is the complete ISASTaskExecution contract:

```
public interface ISASTaskExecution
{
  bool Cancel();
  int ResultCount { get; }
  ISASTaskStream OpenResultStream(int Index);
  RunStatus Run(ISASTaskTextWriter LogWriter);
}
```

Because this special contract is not part of the SAS.Tasks.Toolkit.SasTask implementation, you must specify it in the class declaration. Here is the C# example:

```
public class DS2Datalines : SAS.Tasks.Toolkit.SasTask,
   SAS.Shared.AddIns.ISASTaskExecution
```

Here is the Visual Basic example:

```
Public Class DS2Datalines
      Inherits SasTask
      Implements ISASTaskExecution
```

Both of these class declarations advertise support for ISASTaskExecution. To seal the deal (and to allow your class to compile), you have to actually *implement* the methods and properties that are spelled out in the contract.

Cancel: Support Is Optional

Let's begin with the easy part. The purpose of ISASTaskExecution.Cancel is to allow a user to cancel your task while it is running. In SAS Enterprise Guide, canceling is the operation that occurs when you find the task running in the Task Status view, you right-click on it, and you select **Stop**.

The easiest way to implement the cancel operation (and the way it's done in this task example) is to simply return a false value. This is a way of telling SAS Enterprise Guide, "I'm sorry, this task does not support a cancel operation midstream, just be patient while I finish my work." Here is the C# example:

```
public bool Cancel()
{
   // this task does not implement Cancel, so
   // return false
   return false;
}
```

If you have a task that can result in a very long-running operation, you might want to consider a more accommodating approach. In this approach, you actually cancel the running operation and return a true value, thus clearing the task from its running status. Accomplishing this can be trickier than it seems on the surface. To initiate the cancel operation, you must design your running operation so that it can be interrupted and recovered from a partially completed state.

ResultCount: How Many Results?

The ISASTaskExecution.ResultCount property provides a way for the application to ask your task, "How many discrete pieces of results will you create?" For this task example, the only result that it creates is a SAS program, so the answer to this question is always going to be 1. Here is the C# example:

```
public int ResultCount
{
    get { return 1; }
}
```

For each result that your task claims it creates, SAS Enterprise Guide calls into the task to retrieve a result using the SAS.Shared.AddIns.ITaskStream interface. The SAS Task Toolkit provides a helper class that makes it easy to wrap your results in a structure that implements the ISASTaskStream contract. In this task example, results are created and packaged by ISASTaskExecution.Run.

Run, Task, Run!

The Run method is where all of the important action happens. SAS Enterprise Guide calls ISASTaskExecution.Run when it's time for your task to go to work.

The main work of this task is to read a data set and churn out a DATA step program. With the DatasetConverter class, this requires just a few lines of code:

```
DatasetConverter convert = new DatasetConverter
    (Consumer.ActiveData.Server,
     Consumer.ActiveData.Library,
     Consumer.ActiveData.Member);
string code = convert.GetCompleteSasProgram(OutputData);
```

Now that the SAS program is in the variable named **code**, you have everything that you need to create and package the results. The SAS.Tasks.Toolkit.Helpers.ResultInfo class provides a convenient wrapper for the ITaskStream interface, which enables you to get the task results into SAS Enterprise Guide. Here is the C# example that packages the SAS program into a result that SAS Enterprise Guide can use:

```
// this builds up the result structure.
// It contains the SAS program
// that was constructed.  By tagging it with the
// "application/x-sas" mime type, that serves as a cue
// to SAS Enterprise Guide to treat this as the generated SAS
// program in the project.
resultInfo = new SAS.Tasks.Toolkit.Helpers.ResultInfo();
resultInfo.Bytes = System.Text.Encoding.ASCII.GetBytes(
    code.ToCharArray());
```

```
resultInfo.MimeType = "application/x-sas";
resultInfo.OriginalFileName =
  string.Format("{0}.sas",
  Consumer.ActiveData.Member);
```

This ResultInfo package is unwrapped when SAS Enterprise Guide calls the OpenResultStream method (which is part of the ISASTaskExecution interface). Here is the C# example:

```
public ISASTaskStream OpenResultStream(int index)
{
  // this task returns just one result, the SAS program.
  // So you handle this only when the index is 0.
  if (index == 0)
  {
      return new SAS.Tasks.Toolkit.Helpers.StreamAdapter(
        new MemoryStream(_resultInfo.Bytes),
          resultInfo.MimeType,
          resultInfo.OriginalFileName);
  }
  else return null;
}
```

While the task is running, you can issue status and diagnostic messages to the SAS log using the ISASTaskLogWriter interface.

When SAS generates messages for a log, certain message lines have special meanings and are in different colors. For example, ERROR lines are usually colored in red. SAS.Tasks.Toolkit contains a helper class that you can use to control your log content in a similar way. The class is called SAS.Tasks.Toolkit.Helpers.FormattedLogWriter and it contains static methods to indicate special lines in the log. Here's how this task example uses this class:

```
SAS.Tasks.Toolkit.Helpers.FormattedLogWriter.WriteNormalLine
   (LogWriter, Messages.ConvertedData);
SAS.Tasks.Toolkit.Helpers.FormattedLogWriter.WriteNoteLine
   (LogWriter, string.Format(
      Messages.ConvertedDataMetrics,
      convert.ColumnCount,
      convert.RowCount));
```

When you run this task on the SASHELP.CARS data, the log shows these messages, with the NOTE line colored appropriately:

```
Converted SAS data set to DATA step program.
NOTE: Read 15 columns and 428 rows of data
```

Chapter Summary

This chapter provides some of the advanced techniques that you can use to integrate untraditional processes into your SAS Enterprise Guide projects. By using the ISASTaskExecution contract, you can substitute your own business logic and workflow for the traditional "select options and generate a SAS program" workflow of a SAS task.

This chapter shows how to use .NET data access techniques to read SAS data sources, making use of data content and data set metadata directly in the SAS DICTIONARY tables.

The chapter provides an example of how to reuse the SAS Enhanced Editor in your own task user interface. You learn how to take advantage of helper classes in the SAS Task Toolkit with a minimum amount of code.

Chapter 13: Putting the Squeeze on Your SAS Data Sets

About This Task	**178**
Example Source Files and Information	178
Adapting the Sample	**178**
Refactoring a Macro	178
Compressing the Data Even Further	179
See How You Did: Adding Reporting	180
Wrapping the Sample in a Task	**183**
Designing a User Interface	183
Modeling the Options in a .NET Class	187
Putting It All Together: Running and Repeating the Task	189
Chapter Summary	**191**

Those of you who have been in the computer industry for a while can remember when it was critical to scrimp and save every byte in memory and every sector on disk because storage was scarce in both mediums. But, the new recruits in the computer industry often do not have the same frugality. Machines have fast-growing capacities for disk space and memory, and not everyone sees the incentive to optimize the use of these resources.

However, SAS programmers do. How do I know this? Because on support.sas.com, "Sample 35230: Shrinking character variables to minimum required length" is a popular and highly rated sample. (To see the sample SAS program, visit http://support.sas.com/kb/35/230.html.)

In this chapter, this popular sample program is extended into a custom task for SAS Enterprise Guide. The program is a macro that examines each character variable in the data set, measures the length of the longest value in the data, and then adjusts the data set to shrink the length of each character variable to just the size that is needed to fit the data. For data sets with lots of observations and grossly overallocated storage, this exercise can result in a significant reduction in file size.

About This Task

This task example is named SAS.Tasks.DataSqueezer.dll. It is built and ready to use. This task is an example of how to adapt a proven and popular SAS program into a useful custom task. It requires a simple user interface with a few options. Most of the work in the task is done in the SAS program, which is embedded in the .NET assembly as a file resource.

Example Source Files and Information

.NET language and version	C# and Visual Basic .NET 2.0 (Microsoft Visual Studio 2010)
.NET difficulty	Low
.NET features	Windows Forms
SAS difficulty	Medium
SAS features	DATA step and macro language
Binaries	SAS.Tasks.DataSqueezer.zip
Source code	SAS.Tasks.DataSqueezer_src.zip

Adapting the Sample

The sample on support.sas.com focuses on one aspect of the data size—the length of a character variable. When you allocate the size of a variable using the LENGTH statement, you tell SAS how many bytes of storage to reserve for the values of that variable in each row. If that size is larger than it needs to be, then your SAS data set might occupy more storage space on disk than is necessary.

The sample program examines the data set, down to each value of each character variable. It calculates the minimum length necessary to accommodate the current values in the data. It creates an output data set with all of the data values, but it changes the size allocations for each character variable to be the minimum length that is necessary.

Note: Reducing the lengths of character variables does not cause any data loss for the current data set. However, it *limits* the capacity of each character variable in the output data set in the future. If you expect to add more records to the output data set in the future, and you want to accommodate data values that are longer than those in the current data set, do not run this task to save space. A shorter length for a character variable can truncate a future data value.

Refactoring a Macro

The sample program is already in the form of a SAS macro, which is usually perfect for a custom task. The macro accepts one parameter—the name of the data set to process. This macro creates an output data set with the same name with an underscore appended to it. (It does not replace the original data set.) The macro in the sample is named %CHANGE, and the example call looks like this:

```
%change(work.bigdata);
```

Running this macro results in an output data set named WORK.BIGDATA_, with the lengths of the character variables optimized for the current data values.

Wouldn't it be nice to provide the users of this task with more control so that they can name the output data set anything they want? You can do this by adding a new parameter to the macro to represent the output data set name.

For readability, you should change the name of the macro program to reflect what the program does. After you've done that, the example call looks like this:

```
%squeezeChars(work.bigdata /* in */, work.smaller /* out */);
```

Compressing the Data Even Further

Shrinking your data set is a lot like trying to pack your clothes efficiently in a suitcase. When you pack your suitcase for a big trip and you're trying to make all of your clothing fit into a tight space, you have just a few options:

- Pack fewer clothes. (Or, store fewer records, which is usually not an option.)
- Go on a crash diet so that you can pack the same amount of clothing, but in smaller sizes. (This is similar to reducing the length of variables, but keeping the values intact.)
- Pack all of the clothes that you need in their current sizes, but be prepared to have to sit on the suitcase to make it shut.

SAS offers compression methods that are applied to the entire data set, which is like sitting on the suitcase to make it shut. These compression methods can reduce the storage size on disk, but there is a small overhead cost for reading and writing the data. (Unlike clothing, when you unpack your data from a compressed data set, it does not come out all wrinkled. With data, the price that you pay is that it takes more time to unpack.)

The COMPRESS= data set option enables you to specify a compression algorithm to apply to your output data. Different data forms respond differently to the two algorithms. The SAS documentation for the COMPRESS= option does a great job of explaining the differences. For this task example, you just need to know that you have three options for COMPRESS: NO (no compression), CHAR (run-length encoding), and BINARY (Ross data compression).

With the COMPRESS= option added, the SAS macro program now has an additional parameter—the one for compression. The example call looks like this:

```
%squeezeChars(work.bigdata, work.smaller, compress=NO);
```

The complete revised SAS macro program, embedded in the Visual Studio solution named squeezeChars.sas, looks like this:

```
%macro squeezeChars(dsn, outputDsn, compress=NO);
data _null_;
  set &dsn;
  array qqq(*) _character_;
  call symput('siz',put(dim(qqq),5.-L));
  stop;
run;

data _null_;
  set &dsn end=done;
  array qqq(&siz) _character_;
  array www(&siz.);
  if _n_=1 then
    do i= 1 to dim(www);
      www(i)=0;
    end;
  do i = 1 to &siz.;
    www(i)=max(www(i),length(qqq(i)));
  end;
  retain _all_;
  if done then
    do;
      do i = 1 to &siz.;
        length vvv $50;
        vvv=catx(' ','length',vname(qqq(i)),'$',www(i),';');
        fff=catx(' ','format',vname(qqq(i))||' '||
          compress('$'||put(www(i),3.)||'.;'),' ');
        call symput('lll'||put(i,3.-L),vvv);
        call symput('fff'||put(i,3.-L),fff);
      end;
    end;
run;

data &outputDsn. (compress=&compress.);
  %do i = 1 %to &siz.;
    &&lll&i
      &&fff&i
  %end;
  set &dsn;
run;
%mend;
```

See How You Did: Adding Reporting

If you're going to the trouble of compressing your data sets, you might want to know how effective the compression is. A before and after report would enable you to easily see how much storage is saved by trimming each variable and what the difference is in the overall file size.

The column metadata, including the column lengths, is included as part of the PROC CONTENTS output. (You can also use the DATASETS procedure with the CONTENTS statement.) For a report that shows a good comparison, you don't want to simply use the output automatically generated from PROC CONTENTS. To make the report easier to read and understand, you need to capture the output from PROC CONTENTS for each data set (before and after), and you need to capture the file size.

Although PROC CONTENTS can report on the file size for a data set, it's easy to pluck this information from one of the special DICTIONARY tables in SAS. There is a special DICTIONARY table that includes a series of views in the Sashelp library. The VIEW member that you want to look at is SASHELP.VTABLE, which tracks all table-level metadata for data sets and views in assigned libraries.

To capture variable lengths and the file size for the before and after tables and to merge the results for reporting, this code does the trick:

```
proc contents data=&inLib..&inMem.
   out=_beforeCols noprint;
run;
data _summarySize;
   set sashelp.vtable(
       keep=filesize libname memname
       rename=(filesize=filesizeBefore)
       where=(libname="&inLib" and memname="&inMem"));
run;
proc contents data=&outLib..&outMem.
   out=_afterCols noprint;
run;
/* custom format to help with reporting */
proc format lib=work;
 value typeName
   1 = "Numeric"
   2 = "Character";
run;
data summaryCols;
merge beforeCols(keep=name type length rename=(length=beforeLength type=varType))
       _afterCols(keep=length rename=(length=afterLength));
format varType typeName.;
run;
data summarySize(keep=filesizeBefore filesizeAfter);
   set _summarySize;
   merge sashelp.vtable(
       keep=filesize libname memname
       rename=(filesize=filesizeAfter)
       where=(libname="&outLib" and memname="&outMem"));
run;
```

With these metrics collected, you now have enough information to create an informative report with a few PROC PRINT steps.

```
title "Summary of changes in column length";
title2 "(&inLib..&inMem to &outLib..&outMem)";
proc print data=_summaryCols noobs;
run;
title "Summary of changes in file size (in bytes)";
proc print data=_summarySize noobs;
format filesizeBefore comma12. filesizeAfter comma12.;
run;
title;
```

Figure 13.1 shows an example of a report. As the report shows, the lengths of a few character columns (such as **Category** and **Subcategory**) have been trimmed to save space. Those changes and other compressions resulted in a savings of nearly 700 KB for the data set.

Figure 13.1: Report Showing Compression Results

Summary of changes in column length
(SAMPLE.CANDY_SALES_SUMMARY to WORK.CANDYCOMPRESSED)

NAME	varType	beforeLength	afterLength
Category	Character	10	5
Customer	Numeric	8	8
Date	Numeric	8	8
Discount	Character	8	3
Fiscal_Month_Num	Numeric	3	3
Fiscal_Quarter	Character	6	6
Fiscal_Year	Character	4	4
Name	Character	20	20
OrderID	Numeric	8	8
ProdID	Numeric	8	8
Product	Character	34	32
Region	Character	9	7
Retail_Price	Numeric	8	8
Sale_Amount	Numeric	8	8
Subcategory	Character	13	9
Type	Character	11	9
Units	Numeric	8	8

Summary of changes in file size (in bytes)

filesizeBefore	filesizeAfter
2,703,360	2,015,232

Wrapping the Sample in a Task

With the SAS program in place and working, the remaining steps for creating this custom task are:

- Design and build a simple user interface to accept the options to use in the SAS macro.
- Create a simple model to represent the options in .NET and to store and retrieve the settings using XML. In this way, the settings are remembered in your SAS Enterprise Guide project.
- Package the SAS code in the task and add the logic that assembles the correct program to run based on the stored settings.

To get started, create a new project in Visual Studio. For the task example in this chapter, the shell of the project was created using the custom task Visual Studio project template. For more information, see Chapter 3, "Creating Custom Task Projects in Microsoft Visual Studio."

Designing a User Interface

Even though there is quite a bit going on in the SAS program in this task, there are only a few options for the user to select. The task needs a simple user interface that enables the user to:

- Select the name and location of the output data set (the compressed data set). This information can be captured with a text box and a **Browse** button that allows the user to navigate SAS libraries.
- Specify what level of compression to apply—none, run-length encoding, or Ross data compression. This is best represented by a drop-down combo box, which offers the three valid options in easy-to-read wording.
- Specify whether to create a report that summarizes the compression results. Because this is basically a Boolean option (yes or no or true or false), this can be a simple check box.

Figure 13.2 shows the user interface as it appears in the Visual Studio project. It includes all of the elements and a descriptive text label that helps the user understand the task better before using it.

Figure 13.2: The User Interface in the Design View

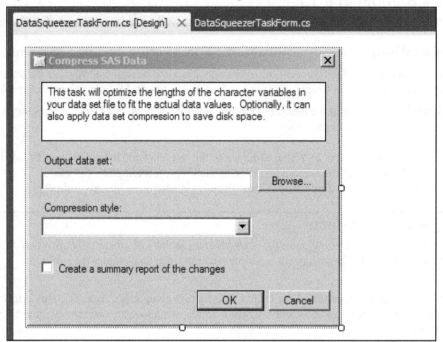

The .NET code behind the UI controls is not complex and you can examine it in the Visual Studio project. In the next section, you will learn about two special aspects of the user interface: handling the **Browse** button action and populating the **Compression style** menu with available options.

Browsing the SAS Environment to Select Output

The File dialog box that enables you to navigate SAS libraries is built into SAS Enterprise Guide. It's not something that you would want to have to build yourself. That's why the consumer API provides a method for you to show that dialog box, enable the user to make a selection from it, and then use that selection in your task.

The .NET event handler for the **Browse** button looks like this (in C#):

```
/// <summary>
/// This is the handler for the Browse button to select an
/// output data set.
/// It makes a single call to the consumer application
/// (SAS Enterprise Guide) to show the output data selector,
/// pinning the output to a single server (same server
/// where the input data resides).
/// </summary>
```

```csharp
/// <param name="sender"></param>
/// <param name="e"></param>
private void btnBrowse_Click(object sender, EventArgs e)
{
  string cookie ="";
  ISASTaskDataName name =
    Consumer.ShowOutputDataSelector(this,
    ServerAccessMode.OneServer,
    Consumer.AssignedServer,
    "", "", ref cookie);

  // if the user closes the dialog without making a selection,
  // the name will be null.
  if (name != null)
  {
    txtOutputData.Text =
      string.Format("{0}.{1}", name.Library, name.Member);
  }
}
```

The call to ISASTaskConsumer.ShowOutputDataSelector triggers the display of the File dialog box. It's like a Save As dialog box for SAS data, allowing you to navigate the SAS server environment and select an output library and member name. If the user selects a member name that already exists, the dialog box prompts the user with a "Do you want to overwrite?" question. All of this action is built into the consumer API and works the same way whether your task runs in SAS Enterprise Guide or in the SAS Add-In for Microsoft Office.

Validating User Input

For convenience, this task's user interface enables the user to enter a name for the output data set, rather than browsing for a location. Some users prefer the point-and-click approach, while other users want to enter the name of a known location. If you allow direct entry into a text field, then you must provide validation to ensure that the value entered by the user complies with the rules of your task.

In this example, the task needs the output data set name to be expressed as *libref.member*. This is what is expected by the SAS program that will eventually run with these settings.

As a task designer, you need to decide how much validation to enforce in your task versus what the user can handle when the resulting program is run in SAS. For example, you could allow the user to enter OUTLIB.MY**DATA in the text field. Keep in mind, the asterisk is not a valid character in the output data set name. As a result, the user will get a syntax error message in the SAS log when the task is run in SAS.

Deciding how much validation to enforce is a balancing act. You want to prevent the user from encountering unnecessary errors, but you want to provide flexibility for more experienced users to make their own selections. (In this way, creating a custom task is like raising a child—you must decide when it's OK for a child to have a little freedom and risk making a mistake.)

In this task example, there is only basic validation of the text field for the output data set name. Validation is triggered when the dialog box is closed after the user clicks **OK**. The validation logic checks the following:

- A value is specified.
- The value is a two-level value—two parts separated by a period.
- Each part contains only characters that are considered valid in a filename (no spaces or special characters are allowed).

If the output data set name does not pass muster, a message is displayed and the operation of closing the dialog box is canceled. The user gets a chance to correct the error.

The validation logic *does not* check the following:

- The output library that is specified actually exists and is assigned.
- The length of each level (libref and member) complies with SAS syntax rules.

If the name passes validation, the value is accepted as the dialog box closes. Beyond that, if the name is not valid in SAS for some reason, the user gets an error message in the SAS log as the task is run.

The following code shows the validation as it occurs in the OnClosing method. This code overrides the default behavior of the OnClosing method in a Windows Form class.

```
// save any values from the dialog into the settings class
protected override void OnClosing(CancelEventArgs e)
{
  if (this.DialogResult == DialogResult.OK)
  {
    // validate data set name - some crude validation
    // a regular expression would be better
    string[] parts =
     txtOutputData.Text.Trim().Split(new char[] { '.' });
    if (txtOutputData.Text.Trim().Contains(" ") ||
       parts.Length != 2 ||
       parts[0].IndexOfAny(
         System.IO.Path.GetInvalidFileNameChars()) > -1 ||
       parts[1].IndexOfAny(
         System.IO.Path.GetInvalidFileNameChars()) > -1)
    {
      MessageBox.Show(
   "Output data name must use valid LIBNAME.MEMBER notation.",
   "Output is invalid");
      e.Cancel = true;
    }
    else
    {
```

```
        Settings.OutputData = txtOutputData.Text;
        Settings.IncludeReport = chkCreateReport.Checked;
        CompressOptions sel =
          cmbCompression.SelectedItem as CompressOptions;
        Settings.Compress = sel.Syntax;
      }
    }
    base.OnClosing(e);
}
```

Modeling the Options in a .NET Class

The SAS syntax for the COMPRESS= option allows one of three values: NO, CHAR, or BINARY. These syntax keywords are not very descriptive, so you should make them easier to understand in the task's user interface.

In this case, it is more convenient to create a .NET class that encapsulates the SAS syntax and the more user-friendly descriptions of the COMPRESS options. You can use this class to help control the business rules that you want to build around the raw SAS syntax.

In this task example, this helper class is named CompressOptions. It's a small class. Here it is in its entirety in C#:

```
/// <summary>
/// Simple class to represent the COMPRESS options and their
/// user-friendly descriptions
/// </summary>
public class CompressOptions
{
  /// <summary>
  /// SAS keyword for the COMPRESS option
  /// </summary>
  public string Syntax { get; set; }
  /// <summary>
  /// Descriptive name for the type of compression
  /// </summary>
  public string Name { get; set; }
  /// <summary>
  /// Return a list of the possible options
  /// for use in a UI element, such as a drop-down
  /// combo box.
  /// </summary>
  /// <returns></returns>
  public static Dictionary<string, CompressOptions>
    GetSupportedOptions()
  {
    Dictionary<string, CompressOptions> opts =
      new Dictionary<string, CompressOptions>();
```

```
            opts.Add("NO", new CompressOptions()
                { Syntax = "NO",
                Name = "None" });
            opts.Add("CHAR", new CompressOptions()
                { Syntax = "CHAR",
                Name = "Run Length Encoding (RLE)" });
            opts.Add("BINARY", new CompressOptions()
                { Syntax = "BINARY",
                Name = "Ross Data Compression (RDC)" });
            return opts;
        }
    }
```

The CompressOptions class represents a simple object: a COMPRESS= syntax value combined with its more readable descriptions. It includes a static helper method that returns all three valid COMPRESS values.

This allows the UI class to take advantage of .NET data binding. .NET data binding is the ability to take a collection of .NET data structures and populate the values in a UI control without having to write code that sets every aspect of the control.

For example, when initializing the drop-down combo box in the dialog box, the code is simple:

```
        /// <summary>
        /// declaration of helper to populate compression options
        /// </summary>
        private Dictionary<string, CompressOptions> availOpts =
            CompressOptions.GetSupportedOptions();

        // init compression options
        cmbCompression.DisplayMember = "Name";
        foreach (string key in availOpts.Keys)
        {
            cmbCompression.Items.Add(availOpts[key]);
        }
```

With these few lines of code (located in DataSqueezerTaskForm.cs), the combo box is populated with all of the valid options, as shown in Figure 13.3. The **DisplayMember** setting uses the **Name** field, which contains the more understandable wording. Now the control contains the objects that serve two purposes: there is a "friendly" name for use in the display, and there is a "syntax" name for use when generating the SAS program.

Figure 13.3: Combo Box Populated Using CompressOptions Class

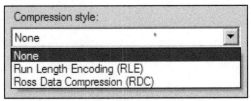

This is a simple example of pairing a single object (a keyword in this case) with a user option. In a complex task with several options, you might have complex objects that result in more permutations of SAS syntax.

Putting It All Together: Running and Repeating the Task

Like many of the task examples in this book, this task uses a settings class to encapsulate task serialization and SAS program generation. (Remember, task serialization is saving and restoring the user-selected options.)

The DataSqueezerTaskSettings class (in DataSqueezerTaskSettings.cs) has members that represent the following task properties:

- compression type
- output data set name
- whether to create a report

The class contains the logic to convert these properties into XML and convert them from XML (ToXml and FromXml methods). It has a method named GetSasProgram that (you guessed it) creates a SAS program based on the settings in the task.

When it comes to storing a task state as XML, there is merit in a simple approach. For example, the code to store the task state looks like this:

```
public string ToXml()
{
  XmlDocument doc = new XmlDocument();
  XmlElement el = doc.CreateElement("DataSqueezerTask");
  el.SetAttribute("outputData", OutputData);
  el.SetAttribute("compress", Compress);
  el.SetAttribute("includeReport",
    XmlConvert.ToString(IncludeReport));
  doc.AppendChild(el);
  return doc.OuterXml;
}
```

This yields a simple XML representation:

```
<DataSqueezerTask
  outputData="WORK.output"
  compress="CHAR"
  includeReport="true" />
```

That's all the information that you need to generate a SAS program using the SAS macro program and other statements described earlier in this chapter.

To make it easier to maintain, the SAS macro program and reporting code is stored in the .NET assembly as embedded resources. This enables you to maintain the SAS program logic in separate files, rather than in-line in the .NET code.

Figure 13.4 shows the structure of the Visual Studio project with the SAS macro program and reporting code stored as embedded resources.

Figure 13.4: Structure of the Visual Studio Project

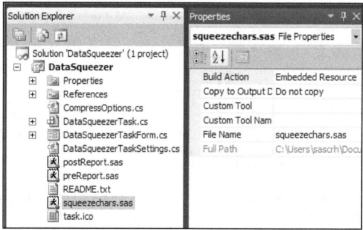

The SAS Task Toolkit library provides a helper method to extract the contents of the file into a **String** variable, which makes it easier to incorporate the stored code into the .NET logic that assembles the complete SAS program the task generates.

```
string macro =
SAS.Tasks.Toolkit.Helpers.UtilityFunctions.ReadFileFromAssembly
   ("SAS.Tasks.DataSqueezer.squeezechars.sas");
```

All of the logic for assembling the SAS program is in the GetSasProgram method in the DataSqueezerTaskSettings.cs file.

Chapter Summary

The exercise in this chapter creates a useful task that compresses the size of a SAS data set. But, more importantly, it shows how simple it can be to convert a SAS program (especially a program that is expressed as a SAS macro) into a custom task.

A proven SAS program provides a solid foundation for a custom task. When you add a user interface to the mix, you can make that task more accessible to a wider variety of users.

Chapter 14: Take Command with System Commands

About This Task	**193**
Example Source Files and Information	194
Building a Task That Runs Commands	**194**
The Structure of This Task	195
Implementing a Task with ISASTaskExecution	196
Chapter Summary	**201**

When experienced SAS users use SAS Enterprise Guide for the first time, they are sometimes surprised by some SAS features that no longer work how they used to. For example, SAS programs that used to escape out to the system shell to run system commands no longer work that way.

This isn't necessarily a problem with SAS Enterprise Guide. It is a by-product of the topology of a client/server architecture. When you run your processes in SAS Enterprise Guide, your system shell might not be on the same machine as your SAS process. Even if you have everything on the same machine, the default configuration does not allow your SAS process to launch other processes using the **x** command or SYSTASK statement.

But, that's okay. You can fill that gap with a simple custom task.

About This Task

This task example is a process that doesn't create or run a SAS program. In fact, it doesn't interact with your SAS process at all. It's conceptually equivalent to a Windows batch file that runs at an appointed location in your SAS Enterprise Guide process flow. You can use it to copy files, create directories, or even launch another process.

The task implements the ISASTaskExecution interface, which tells the host application that it will run itself.

Figure 14.1 shows an example of this task in a process flow. In this example, the task uses the Windows `DIR` command to create a text file with a directory listing of files in a particular folder. The resulting text file is imported into a SAS data set using the Import Data task.

Figure 14.1: Example of the Task in Action

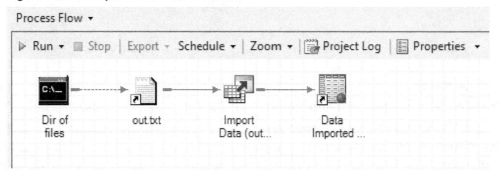

Example Source Files and Information

This task example is named SAS.Tasks.Examples.SysCommand.dll (C# version) and SAS.Tasks.Examples.SysCommandVB.dll (Visual Basic .NET version). They are both built and ready to use.

.NET language and version	C# and Visual Basic .NET 2.0 (Microsoft Visual Studio 2008 and later)
.NET difficulty	Medium
.NET features	Windows Forms, ISASTaskExecution processing, and System.Diagnostics.Process execution
SAS difficulty	Low
SAS features	None!
Binaries	SysCommand_Binaries.zip SysCommandVB_Binaries.zip
Source code	SysCommand_Src.zip SysCommandVB_Src.zip

Building a Task That Runs Commands

The task example in this chapter presents a useful utility. The utility replaces *some* of what SAS programmers traditionally expect when they use the SYSTASK statement or **x** command.

But, it can't replace those constructs completely. The difference between this task and those SAS constructs is this—this task runs in the context of your local PC on your Windows operating system; the **x** command and SYSTASK statement run in your SAS program on the operating system of the SAS session. These two constructs might also be on your local PC, but they could just as well be on a remote server, such as a UNIX system.

You can extend this task by using the concepts that you learn from this example. For example, instead of building a simple batch file (as shown in this example), you can build a sophisticated script that does more by interacting with a remote shell (rsh) or using the Microsoft Windows PowerShell capabilities. (For more information about PowerShell, visit http://technet.microsoft.com/scriptcenter.)

The Structure of This Task

This task is deceptively simple. It has only a few moving parts.

- A simple user interface with a multi-line text box to accept a list of commands to run, as shown in Figure 14.2.
- Logic to save the state (the commands) and reload the state when it's time to run. (This is in most custom tasks.)
- The routine to actually execute the commands and collect any output in the log for the project.

Unlike traditional custom tasks, this task doesn't have to generate a SAS program. Instead of a SAS program that SAS Enterprise Guide runs on your behalf, the logic for running the task is in the task code itself.

Figure 14.2: User Interface of the System Command Task

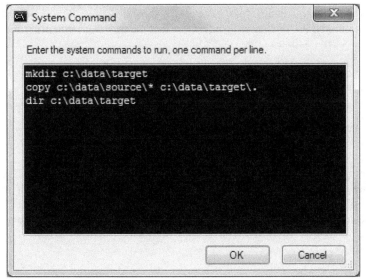

Implementing a Task with ISASTaskExecution

Like most of the task examples in this book, this task example uses the SAS Task Toolkit library to make it easier to implement a task with fewer lines of code. The task class, implemented in SysCommandVB.vb in the Visual Basic version, inherits from SAS.Tasks.Toolkit.SasTask. SAS.Tasks.Toolkit.SasTask handles most of the mechanics of fitting into the application as a task.

To use ISASTaskExecution, you must implement a few additional methods and properties that are not included in SAS.Tasks.Toolkit.SasTask. This requires the following steps:

- Declare your intention to use ISASTaskExecution in the task declaration.
- Implement the required members (methods and properties) to fit ISASTaskExecution. Four members are required (Run, ResultCount, OpenResultStream, and Cancel). For this task, only the Run method requires any substantial content.

Extending a SasTask Class to Include ISASTaskExecution

The only way that SAS Enterprise Guide knows that you're taking control with ISASTaskExecution is the fact that you've implemented that interface in your task class. In Visual Basic, you declare your intention to do this using the **Implements** keyword with **ISASTaskExecution**. Here's an example:

```
<ClassId("f5917eb4-410a-43a5-8f6d-eb02033db083")> _
<IconLocation("SAS.Tasks.Examples.SysCommand.SysCommand.ico")>
<InputRequired(InputResourceType.None)> _
<Version(4.2)> _
Public Class SysCommandVB : Inherits SAS.Tasks.Toolkit.SasTask
    Implements ISASTaskExecution
```

In C#, the idea is the same, but the syntax is different. Instead of a separate keyword, you add the interface name to a comma-delimited list of classes or interfaces that the task class inherits from. Here's an example:

```
[ClassId("E3B71B12-9930-45de-B396-ACF8661E0F48")]
[Version(4.2)]
[InputRequired(InputResourceType.None)]
[IconLocation("SAS.Tasks.Examples.SysCommand.SysCommand.ico")]
public class SysCommand : SAS.Tasks.Toolkit.SasTask,
    SAS.Shared.AddIns.ISASTaskExecution
```

Regardless of the programming language, when you use Visual Studio to add these declarations for you, it automatically adds default implementations of the required properties and methods that are needed. Of course, the implementations won't represent any useful work. In fact, each of these default implementations does nothing except throw a **NotImplementedException**. Here's an example (in Visual Basic):

```
Public Function Run(ByVal LogWriter As ISASTaskTextWriter) _
  As RunStatus _
  Implements ISASTaskExecution.Run

    Throw New NotImplementedException()

End Function
```

This default implementation for each member enables your SAS custom task class to remain buildable. In other words, Visual Studio enables your task to compile and build a DLL file.

Even though it's buildable, your custom task isn't usable, *yet*. The **NotImplementedException** exception stops your task dead in its tracks when you try to run it. For the task example in this chapter, not all of the ISASTaskExecution required members need full implementations. Here are brief descriptions of the treatments that they do need:

- Cancel method: Return False, indicating that Cancel is not supported when the task is running. In a sophisticated implementation, you could add logic to allow the work in the Run method to be interrupted. However, you would have to schedule that work in a separate Windows thread, which adds a layer of complexity to the task. For long-running tasks that you want to allow the user to interrupt, this can be a worthwhile exercise.
- ResultCount property: Return the integer 0. Outside of the log, there are no results to return to the project.
- OpenResultStream method: Keep NotImplementedException in place. Because the task returns 0 for ResultCount, this method is never called.
- Run method: This is where the logic for running the system commands goes. This method represents the real business logic of the task.

Remember, this task doesn't create or run a SAS program. SAS Enterprise Guide needs to know that to behave accordingly, so there is one more property that you need to set in the SasTask class. GeneratesSasCode should be set to False. You can change the property setting in the InitializeComponent method or change it in the Design view, as shown in Figure 14.3.

Figure 14.3: Design View of Properties for the SysCommandVB Task

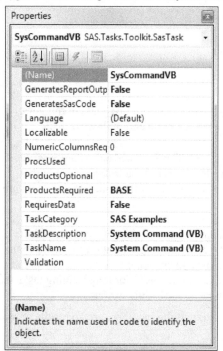

Don't Forget: Writing Output to the Log

SAS users are accustomed to detailed information in the SAS log. The ISASTaskExecution interface puts *you* in charge of putting relevant content in the SAS log when the task runs. In addition to plain text messages, you can decorate your log output with NOTE, WARNING, and ERROR messages that get the same color treatment that they would for SAS program output.

To format the content in the log, use the SAS.Tasks.Toolkit.Helpers.FormattedLogWriter class. For example, to write a NOTE line that indicates which machine you're running the task on, use code like this (in Visual Basic):

```
Imports SAS.Tasks.Toolkit.Helpers
' the Imports statement allows for shorthand use of the
' class name without the full namespace later
FormattedLogWriter.WriteNoteLine(LogWriter, _
    String.Format("NOTE: Running system commands on {0}." & _
        vbNewLine & "Output:", machineName))
```

You should carefully consider the types of messages that users will find helpful. Consider the messages that are important for normal successful operation. Consider the conditions that might merit a WARNING message. And, of course, if the task encounters conditions that prevent it from completing properly, be sure to include an informative ERROR message.

Completing the Run Method

Here is the complete Visual Basic implementation of the Run method:

```
Public Function Run(ByVal LogWriter As ISASTaskTextWriter) _
   As RunStatus Implements ISASTaskExecution.Run
   Dim rc As RunStatus = RunStatus.Success
   ' to keep track of elapsed time for the system commands
   Dim start As DateTime = Now
   ' seed the machine name for use in the log
   ' you need to make it clear that the system commands
   ' are executed on the local machine, not on a
   ' remote SAS server machine.
   Dim machineName As String = "local machine"
   Try
        ' Environment.MachineName can throw an
        ' InvalidOperationException
        machineName = Environment.MachineName
   Catch ex As Exception
      ' couldn't get the machine name
   End Try


   ' the FormattedLogWriter helps color-code your log output
   ' for NOTE, ERROR, and WARNING lines.
   FormattedLogWriter.WriteNoteLine(LogWriter, _
      String.Format("NOTE: Running system commands on {0}." & _
         vbNewLine & "Output:", machineName))
   Try
        Dim log As String = ExecuteCommands()

        ' write the output collected from stdout
        FormattedLogWriter.WriteNormalLine(LogWriter, log)
   Catch ex As Exception
      ' if there is an error, place it in the log
        FormattedLogWriter.WriteErrorLine(LogWriter, _
        String.Format("ERROR: Could not run commands " & _
           vbNewLine & "{0}", ex.Message))

        ' return error status so that it gets the
        ' "red X" treatment
        rc = RunStatus.Error
   End Try

   Dim elapsedTime As TimeSpan = _
      TimeSpan.FromTicks(Now.Ticks - start.Ticks)
```

```vb
        FormattedLogWriter.WriteNoteLine(LogWriter, _
          String.Format("NOTE: System commands completed." & _
          vbNewLine & vbTab & "Real time: {0:F2} seconds", _
          elapsedTime.TotalSeconds))
        Return rc
    End Function
```

If you read the code carefully, you might have noticed that the real meat of the work—running the commands—is shunted to another method named ExecuteCommands. That method uses System.Diagnostics.Process to spin off another process in the Windows environment. Here is the Visual Basic implementation of the ExecuteCommands method:

```vb
    Public Function ExecuteCommands() As String
        Dim stdout As String

        ' create a temp batch file to run the commands
        Dim fn As String = Path.GetTempFileName()
        File.Move(fn, fn & ".cmd")
        fn += ".cmd"

        Dim sw As StreamWriter = New StreamWriter(fn)
        sw.Write(cmds)
        sw.Dispose()

        ' launch a process to run the batch file
        ' record the stdout so you can report in the log for the task
        Dim psi As System.Diagnostics.ProcessStartInfo = _
            New System.Diagnostics.ProcessStartInfo()
        psi.FileName = fn
        psi.RedirectStandardOutput = True
        psi.UseShellExecute = False
        psi.CreateNoWindow = True

        Dim p As System.Diagnostics.Process = _
            System.Diagnostics.Process.Start(psi)
        stdout = p.StandardOutput.ReadToEnd()

        ' does not return control until the job is done
        p.WaitForExit()

        ' clean up the temp file
        File.Delete(fn)

        ' return the stdout
        Return stdout
    End Function
```

The System.Diagnostics.ProcessStartInfo settings determine how and when control returns to the calling process (your task in SAS Enterprise Guide). If you are familiar with the XWAIT and XSYNC system options in SAS, you might recognize these settings. In concept, they are the same mechanisms that SAS uses to return control to your SAS session when you run an **x** command or SYSTASK statement.

A Note of Caution about ISASTaskExecution

Running with ISASTaskExecution is like running with scissors. It can be a little bit dangerous.

When you implement the ISASTaskExecution interface in your custom task, SAS Enterprise Guide puts you in the driver's seat. While running a project or process flow, when it is time to run your custom task, complete control is handed over to your task code by the ISASTaskExecution.Run method. If anything bad happens in that Run method (such as an unhandled exception), it can put the host application (in this case, SAS Enterprise Guide) in a bad state. The user can lose work. Causing the user to lose work is one of the most serious infractions that you can commit as a software developer. So, it's best to be extremely cautious.

For more information about code stability and debugging techniques, see Chapter 8, "Debugging Techniques: Yes, You Will Need Them."

Chapter Summary

In this chapter, you learned that custom tasks are not limited to generating SAS programs. And, a custom task can be very useful for filling a gap in a system function in SAS Enterprise Guide.

The ISASTaskExecution interface enables you to add almost any process, whether it is based in SAS or some other system, to a SAS Enterprise Guide process flow.

Chapter 15: Running PROCs on Your Facebook Friends

Facebook to SAS: The Approach	**204**
Example Source Files and Information	205
Gathering Data from Facebook	**205**
Example of Transforming JSON to DATA Step Statements	205
Analyzing Data from Facebook	**206**
Preparing Data for Reporting	207
Creating Reports That Provide Insight	208
Using the Facebook API	209
Running the Example	**210**
Inside the Task	**211**
Connecting to Facebook and Collecting Data	212
Modeling Data Records with .NET Data Structures	213
Keeping the User Interface Responsive	216
Saving the Results in Your SAS Enterprise Guide Task	217
Chapter Summary	**218**

The explosion of social media sites in recent years has transformed how people communicate with each other, when interacting personally or promoting businesses. At the heart of social media sites is data, structured and unstructured. Wouldn't it be interesting to access that data from within SAS, where you can analyze it and report on it easily?

This chapter presents a custom task that connects to Facebook, one of the most popular social media sites. Using a .NET implementation of the Facebook API, you can collect data from your Facebook profile in a structured form, and then turn it into a series of SAS data sets. Once you've got the data in SAS data sets, you have an entire world of SAS power that you can apply to it.

The approach that this task example demonstrates can be used for other social media sites as well. For example, the microblogging platform Twitter can provide lots of information about Twitter relationships and tweet content via its API.

Facebook to SAS: The Approach

Currently, Facebook is the largest and most extensible social media platform on the planet. With more than 750 million users around the world using Facebook as a platform to connect to one another, it's no wonder that there have been tens of thousands of applications created to help Facebook users interact with each other. With that many applications out there, created by everyone from professionals to high school students, it seems like it wouldn't be *that* difficult to build one, right?

This custom task in this chapter is, in part, a Facebook application. The small part of the task that is a Facebook application connects to Facebook and collects data. The interesting part of the task example—the part that produces a cool SAS report—makes use of good old-fashioned SAS programming. Here is the general approach that you'll follow in this task:

- Use the Facebook API to connect to a Facebook account. Download the friend user data in JSON (JavaScript Object Notation) form. (JSON is a standard transport format for web-based APIs.) You'll use a .NET-based SDK for Facebook to make this step even easier.
- Use a JSON parsing library to transform the user data into SAS data sets.
- Use SAS procedures to create a report of user status messages. Analyze a few of the numerical data points that can be gleaned from basic user profile information.

Example Source Files and Information

The task example is named SasFacebookTask.dll. Because this example also depends upon other third-party DLL files, there are special considerations for how to deploy this task. See **Running the Example** later in this chapter for more information.

Here are some details about the task example:

.NET language and version	C# and Visual Basic .NET 3.5 (Microsoft Visual Studio 2010)
.NET difficulty	Medium
.NET features	Windows Forms, simple menus, third-party SDK (software development kit), asynchronous API calls, and using a web service
SAS difficulty	Medium
SAS features	DATA step, FORMAT procedure, SGPLOT procedure, and REPORT procedure
Binaries	SASPress.Facebook.zip
Source code	SASPress.Facebook_src.zip

Gathering Data from Facebook

In this custom task, an application uses Microsoft .NET to connect to Facebook. (The connection is authenticated with OAuth, an open standard for web authorization.) The application retrieves data from your Facebook profile using the Facebook API, and then produces a complete SAS program. The application can be used in SAS Enterprise Guide 4.2, 4.3, or 5.1. In addition, it can be used as a stand-alone application.

The general steps for developing this Facebook application are:

1. Log on to the Facebook site and register as a developer.

    Use the Facebook developer website at http://developers.facebook.com/ to register this new Facebook application.

    The main requirement is to give your application a name. In return, Facebook assigns an application key (a unique 32-character identifier) to your application. When you build your application, you will use this key in the API calls so that Facebook and the user will know which application is using Facebook. As a result, Facebook is able to track the use of your application.
2. Develop your application using the technology of your choice. For this example, you are using Microsoft .NET.

Facebook supports a few different styles of APIs. The most popular and the most flexible style is called the Graph API, which enables you to send API requests for specific data using HTTP. The responses to these API requests are returned as structured data using JSON. Interpreting JSON requires the use of special development libraries, which are readily available for many popular development technologies. Alas, there is not currently a way to parse JSON with the SAS language so you must transform the JSON results into a form that SAS can interpret.

Your new Facebook application transforms the JSON results into a SAS program with DATALINES statements that contain the data records.

Example of Transforming JSON to DATA Step Statements

Here's an example of a JSON response that the application receives when requesting information about a user's education history:

```
"id" : "21801765",
"schools":
{[
   {
     "school": {
        "id": "110451125642042",
        "name": "Chenango Forks High School"
     },
```

```
      "type": "High School"
    },
    {
      "school": {
        "id": "103788109660536",
        "name": "SUNY Cortland"
      },
      "year": {
        "id": "121127321230671",
        "name": "2006"
      },
      "type": "College"
    }
  ]}
```

With your Facebook application, you can parse the JSON response and transform it into SAS DATA step statements with equivalent data lines:

```
data schools;
length
  UserId $ 15
  SchoolId $ 15
  Name $ 50
  Type $ 20
  Year 8
  ;
infile datalines4 dsd;
input UserId SchoolId Name Type Year;
datalines4;
"21801765","110451125642042","Chenango Forks HS","High School",
"21801765","103788109660536","SUNY Cortland","College",2006
;;;;
```

Analyzing Data from Facebook

Your Facebook application can connect to Facebook and gather the following details from your Facebook profile:

- A list of your Facebook friends.
- For each friend, a few details, such as birthday, education history, and gender.
- A list of recent status messages.

These details are arranged in a SAS program that, when run, yields several data sets that are ready for analysis. The data sets (tables) include:

- WORK.FRIENDS, which contains the friend's full name and an internal ID that Facebook uses. This internal ID is a useful key to join this table with other tables.
- WORK.FRIENDDETAILS, which contains more details about each friend, including first and last name, gender, birthday, URL for the Facebook profile page, and relationship status (for example, Married, Single, and so on).
- WORK.STATUS, which contains a list of recent status messages (within the past two weeks) shared by your friends.
- WORK.SCHOOLS, which contains a list of schools that your friends have attended, including the school name, type (High School, College, Graduate School), and the year graduated.

Preparing Data for Reporting

Each data set contains a column named UserId, which represents the unique ID for a Facebook user. You can use this field as a key to help combine tables that have one-to-many relationships. For example, because each Facebook user might have listed more than one school in his or her education history, you can combine the FRIENDS table and the SCHOOLS table with the following PROC SQL code:

```
proc sql;
   create table work.schoolfriends as
   select t2.name,
          t1.name as schoolname,
          t1.type,
          t1.year
      from work.schools t1 left join work.friends t2
          on (t1.userid = t2.userid)
      order by t1.schoolid;
quit;
```

Using the birthday-related fields in the FRIENDDETAILS table, you can attempt to calculate the calendar birthdays for Facebook users and calculate their ages. However, because many Facebook users do not share their complete birthday information (including birth year), you have to account for missing values. The following program creates a new table with calendar birthdays (formatted as DATE5. values to omit the years from displaying) and ages (if possible):

```
data birthdays (drop=year);
  set friendDetails
      (keep=LastName FirstName
       Gender BirthdayYear BirthdayDay BirthdayMonth);
  length Birthday 8 Age 8;
  format Birthday date5.;
  format Age 6.2;
  if BirthdayYear = . then
```

```
    year = 1899; /* placeholder */
  else year=BirthdayYear;
  Birthday = MDY(BirthdayMonth, BirthdayDay, year);
  if BirthdayYear = . then
    Age = .;
  else Age = yrdif(Birthday, today(), 'act/act');
run;
```

Creating Reports That Provide Insight

The custom task including your Facebook application can produce basic report output, including frequency tables and charts of friends by gender and relationship status, lists of birthdays, and statistics on calculated ages. However, you must be careful when drawing conclusions about your entire population of friends because many Facebook users do not share all of their personal information.

Determining Gender Distribution

For example, Figure 15.1 shows a frequency chart created with PROC FREQ and ODS GRAPHICS. It illustrates the breakdown of gender for one set of Facebook users. At first glance, it looks like there are slightly more males than females in this group.

Figure 15.1: Distribution of Gender for One Set of Friends

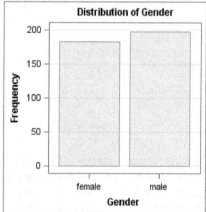

Determining the Ages of Your Friends

One of the more popular social interactions on Facebook is the birthday message. Although most Facebook users share some of their birthday information with friends, many choose to not share their birth years. The data collected by your Facebook application includes birthday information. For those friends that share their birth years, you can calculate their current ages and summarize them, as shown in Figure 15.2.

Figure 15.2: PROC MEANS Output of Known Ages

Analysis of Friend published ages

Analysis Variable : Age

Minimum	Maximum	Mean	Median	N	N Miss
17.4692717	101.9679018	45.5084174	44.2431694	118	266

As you can tell by the **N Miss** field in the report, most of these Facebook users do not share detailed birthday information (only 118 out of 384, in this case).

Because Facebook is often used by people who want to connect or reconnect with classmates from high school or college, many Facebook users *do* share their education histories. The education history includes the names of the schools that a person attended and the years that they graduated. By collecting the education history and making an assumption about the likely age that a person would be when he or she graduated from high school or college, you can estimate the age of each friend and summarize accordingly. Figure 15.3 shows a new report that takes education histories into account.

Figure 15.3: Estimated Ages Based on Education Histories

Calculated Ages based on Graduation years

Analysis Variable : age

How determined?	N Obs	Minimum	Maximum	Mean	Median	N
College	185	19.0000000	74.0000000	44.0216216	44.0000000	185
High School	209	17.0000000	74.0000000	45.1148325	44.0000000	209
Published	118	17.4692717	101.9679018	45.5084174	44.2431694	118

The **N Obs** field shows how many observations contained nonmissing values for the calculated ages. By inferring an estimated age using education history, you can consider more data records. However, it's possible to calculate the wrong age for a person who didn't follow the typical pattern of graduating high school at age 18 or college at age 22.

Using the Facebook API

Here's another thing that Facebook is famous for—changing. Just when you are accustomed to interacting with Facebook in a certain way, it seems that the folks at Facebook find some way to *improve* your experience by changing it out from under you.

As a software developer, I completely understand their point of view. They don't want their product to stand still and get stagnant. However, sometimes their changes affect the Facebook API as well, and that can make it challenging to maintain a Facebook application.

The Facebook service uses an API that is based on REST (representation state transfer) architecture, which is common among Internet-based web services. Your Facebook application communicates with the Facebook service using HTTP, just like a web browser.

For the Facebook application in this custom task, I used a free Facebook SDK that is maintained by another group of developers. The advantage of an SDK is that you don't need to understand every nuance of the Facebook API. The details are abstracted for you, and you can easily code the two simple operations that you need to perform: connect to Facebook and collect the data. It's very similar to the SAS Task Toolkit described in this book. Using an SDK removes the tedious part of the process and enables you to concentrate on your content.

The SDK that I used is called Facebook C# SDK. It was written especially for .NET developers and is available for free at http://csharpsdk.org/. You can download the SDK and even the source code from that location.

The SDK is packaged in a series of .NET assemblies (DLL files). Because the task example makes direct use of those .NET assemblies, you need to provide the assemblies with the finished task. The license agreement for the Facebook C# SDK allows this.

Running the Example

The pre-built task example is available in the **Binaries** location listed in the **Example Source Files and Information** section. The instructions for how to deploy and run the task are included in the task package. (See the Readme file.) Here are some special considerations to note:

- Because the task uses the Facebook C# SDK, there are additional .NET assemblies to deploy. In addition to the task DLL file, there are a few DLL files that come as part of the SDK. These DLL files contain the functions necessary to connect to Facebook and display a login dialog box so that you can sign in to your Facebook profile.
- These DLL files must be kept together when you deploy the task. You can use a special feature of task deployment to help organize the DLL files on your Windows file system. Instead of copying the files loosely into one of the designated **Custom** subfolders (for example, %appdata%\SAS\EnterpriseGuide\4.3\Custom for SAS Enterprise Guide 4.3), you can create an additional folder level to keep these files organized and together.

    For example, you can create a folder named:

    ```
    %appdata%\SAS\EnterpriseGuide\4.3\Custom\Facebook
    ```

    Then, copy the additional DLL files into it. SAS Enterprise Guide still discovers the task automatically, and the subfolder reduces the chance of a filename conflict in the **Custom** subfolder when you have multiple tasks that have dependencies.

Inside the Task

The main steps for building this task include:

- Create a new Visual Studio project (a Class Library project, as described in Chapter 7, "Your First Custom Task Using C#").
- Add references to the Facebook C# SDK assemblies.
- Create a Windows Form that enables you to connect to Facebook and gather data.
- Create a Facebook application that transforms the data for use by SAS, creates a simple report, and adds the data sets to SAS Enterprise Guide for further use.
- Create a .NET class that ties it all together, putting the JSON data and SAS programming logic into one SAS program that can run in SAS Enterprise Guide.

This task example includes a bonus: it can be used as a stand-alone application that can be run *outside* of SAS Enterprise Guide. This enables SAS users who don't have access to SAS Enterprise Guide to experiment with reporting on their data from their Facebook profile.

It's a good example of reuse because there is very little code that is needed to run the task outside of SAS Enterprise Guide. The stand-alone application (which produces a small EXE file) contains a simple call to show the task dialog box. You can then use the menus in the task dialog box to save the Facebook data as a SAS program.

The approach for developing this task is slightly different from other tasks in this book. For other tasks, it's recommended to prototype the results of the task and the SAS program that would generate those results. Once you have the right SAS program, it's easier to create a user interface for a task that produces those results.

However, for this task, you can't prototype the SAS program without having at least a sample of what the data looks like. So, you have to proceed in a different manner:

- Build enough of the task so that it can connect to Facebook and collect a sample of the data in JSON format.
- Use the JSON sample to design a record layout for a SAS DATA step statement so that the data can be read into a SAS data set.
- Design the remainder of the task and put any finishing touches or controls on the user interface that are needed.

Connecting to Facebook and Collecting Data

To connect to Facebook with your Facebook application, you must add references to the two assemblies in the Facebook C# SDK that makes this possible. The assemblies are:

- Facebook.dll
- Newtonsoft.Json.dll (for parsing JSON content)

Add both of these to the task as assemblies. Make sure that **Copy Local** is set to **True** (not **False** as you have done in previous tasks). You need the assemblies to be copied to the output directory and eventually to the location where the task is deployed. These are third-party assemblies and are not already provided in the SAS application.

The task source code contains a simple user interface class—FacebookConnectorForm (located in FacebookConnectorForm.cs). Figure 15.4 shows the task's user interface in Visual Studio in the Design view.

Figure 15.4: The Task User Interface in Visual Studio

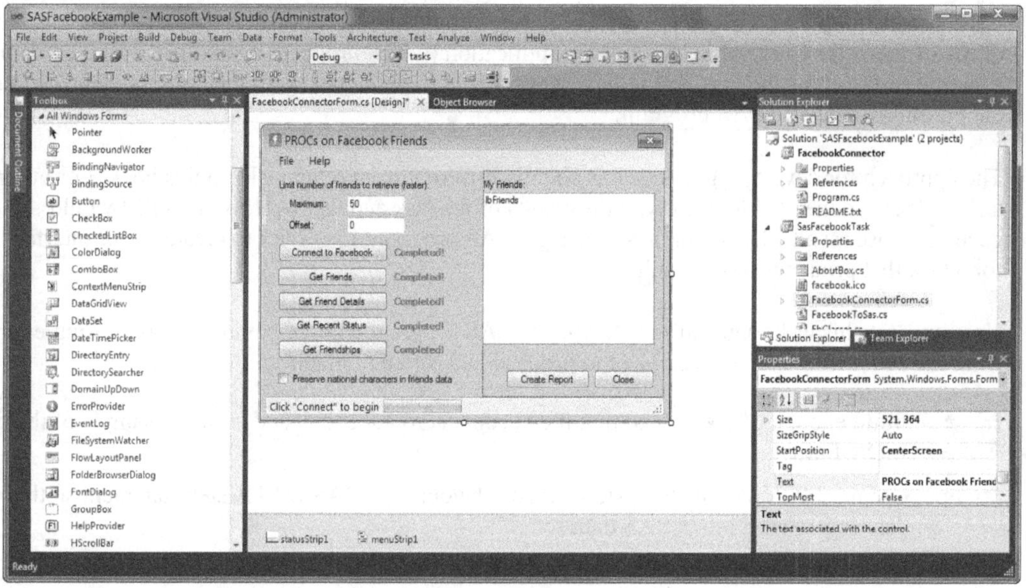

Connecting to Facebook: Get In and Get Out

Some Facebook applications allow a lot of back-and-forth interaction. They allow you to update your Facebook status, upload photos, add comments to the content added by friends, and so on. But, your Facebook application has just a Read-Only view of your friends' data. It connects to Facebook, collects that data, and gets out.

Because gathering data from Facebook can be time-consuming, your Facebook application is designed to collect data in several phases, each of which you initiate by clicking a button. The data-collection phases (and buttons) are:

Connect to Facebook
Pops up a browser window that enables you to log on to Facebook.

Get Friends
Retrieves a list of your friends and their internal Facebook unique IDs. You can throttle the volume of friend data that you collect by specifying a maximum number of friend records to fetch.

Get Friend Details
For each Facebook friend retrieved, this phase calls the Facebook API to collect details: birthday, education history, relationship status, and more. If you fetched 100 friend records in the previous step, this step makes 100 Facebook API calls.

Get Recent Status
For each Facebook friend retrieved, this phase retrieves the most recent status message that was posted in the past week. This API call is intentionally limited to just one status message for each friend in the past seven days. It's possible to fetch more or look back further in time, but it takes longer to retrieve that data.

Get Friendships
This phase can result in a very long operation, so use it with care. It uses the Facebook API to check whether each of your friends is a friend with any of your other Facebook friends. The API method to do this enables you to check only one friend pair at a time, so it results in ($n * (n+1)) / 2$ API calls. For example, for 50 friends, that translates to 1,275 separate API calls!

Modeling Data Records with .NET Data Structures

As you retrieve data from Facebook, you need an intermediate location to hold the data. The data values will ultimately be transformed into a SAS program consisting of a series of DATA step statements.

The key piece of data that the task retrieves is a list of friends. But, it also retrieves data records that you can associate with each friend. Some data records have a one-to-one relationship with each friend record; for example, each friend has just one birthday. Some data records have a one-to-many relationship with each friend record; for example, each friend might have attended multiple schools.

To represent these relationships cleanly, you need a simple *data model*—a method to represent the different data entities and their relationships with each other. Figure 15.5 shows a simple data model in the SAS Enterprise Guide Query Builder. As the figure shows, the **UserId** field is the unique key that enables you to join the different data entities.

Figure 15.5: Friend Data Entities Joined in the Task

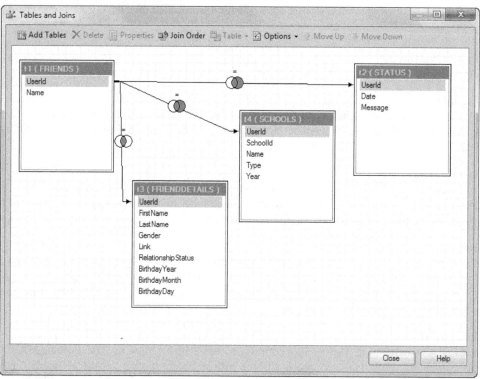

In SAS, you can implement this simple data model with a collection of SAS data sets. In .NET programming, you can implement the data model with simple .NET classes. Here are the .NET classes that the task uses to represent the data retrieved from Facebook. It's a set of three simple .NET classes with properties. Most properties are represented as **String** objects because that's often the easiest way to retain the exact values from Facebook. Once the data is in SAS, it's easy to manipulate using the SAS programming language.

```
public class FbFriend
{
    public string Name { get; set; }
    public string FirstName { get; set; }
    public string LastName { get; set; }
    public string Gender { get; set; }
    public string RelationshipStatus { get; set; }
    public string Link { get; set; }
    public string UserID { get; set; }
    public string BirthdayDay { get; set; }
    public string BirthdayMonth { get; set; }
    public string BirthdayYear { get; set; }
}
```

```
public class FbSchool
{
    public string UserID { get; set; }
    public string SchoolID { get; set; }
    public string Name { get; set; }
    public string Type { get; set; }
    public string Year { get; set; }
}

public class FbStatus
{
    public string UserID { get; set; }
    public DateTime UpdateTime { get; set; }
    public string Status { get; set; }
}
```

These objects can be assembled into collections using the .NET System.Collections.Generic classes:

```
// represent "me" as a Facebook user object
FbFriend me = new FbFriend();

// hashtable of Friend IDs mapped to Friend objects
Dictionary<string, FbFriend> Friends =
    new Dictionary<string, FbFriend>();

// collection of Status objects
List<FbStatus> Statuses = new List<FbStatus>();

// collection of School objects
List<FbSchool> Schools = new List<FbSchool>();

// for keeping a list of "friend pairs",
// tracking connections
List<string> FriendPairs = new List<string>();
```

Turning the Data into a SAS Program

After all of the Facebook data has been gathered, you need to turn the .NET objects into one SAS program. A .NET object is its intermediate format before it is transformed into a SAS data set.

The task example contains a .NET class that is especially devised to make this transformation. The class is aptly named FacebookToSas (located in FacebookToSas.cs).

The FacebookToSas class navigates the collection of .NET objects and turns it into one big SAS program that includes DATA step statements and the basic reporting steps that you want to include in the results. The main method that does this work is called GetSasProgram:

```
static public string GetSasProgram(FbFriend me,
    Dictionary<string, FbFriend> friends,
    List<FbSchool> schools,
    List<FbStatus> statuses,
    List<string> friendPairs,
    bool preserveEncoding)
{ /* see source project */ }
```

Keeping the User Interface Responsive

Some Facebook interactions can take considerable time to complete. Because you want the best user experience possible, it's important to not block the user interface from updating during long-running operations.

The way to achieve this is to use asynchronous programming methods to retrieve the data from Facebook. In .NET programming (and in Windows programming in general), asynchronous programming allows longer operations to run on a different thread. The UI thread remains available for status updates and user interaction.

The Facebook C# SDK provides a set of asynchronous programming methods. When you use an asynchronous programming method, you must provide a callback routine in your code. The callback routine executes when the asynchronous operation is completed.

Because the callback routine can happen on a different Windows thread, you must not attempt to update the user interface until you marshal the code operation back onto the UI thread. If you forget to do that, you might experience a catastrophic "invalid cross-thread operation" exception that can cause everything to vanish.

In this task, marshaling is handled by an anonymous method and the BeginInvoke method, which puts the code operation back onto the UI thread. Here's an excerpt from the Get Friend Details operation:

```
foreach (FbFriend f in Friends.Values)
{
  try
  {
    /* performs work on background thread */
    app.GetAsync(f.UserID, (val) =>
    {
        /* use BeginInvoke to marshal back onto UI thread */
      BeginInvoke(new MethodInvoker(
        delegate()
        {
          if (val != null && val.Result != null)
          {
            /* process values retrieved */
          }
```

```
                    if (tsProgressBar.Value++ ==
                      tsProgressBar.Maximum - 1)
                    {
                           /* update status when complete */
                      tsProgressBar.Visible = false;
                      tsStatusMessage.Text =
                       string.Format(
                         "Details retrieved for {0} friends",
                         Friends.Values.Count);
                      lblStep3.Visible = true;
                    }
                  }
                ));
            });
        }
        catch
        {
          /* update progress bar */
          tsProgressBar.Value++;
        }
      }
```

`BeginInvoke`, `MethodInvoker`, and the `delegate` keyword are all worth learning more about. The details of these are beyond the scope of this book. You can learn about them from msdn.microsoft.com and many .NET community websites (such as www.codeproject.com).

Saving the Results in Your SAS Enterprise Guide Task

In this task, the bulk of the work happens while the task user interface is shown. That's when you connect to Facebook and gather data to include in the data model.

Because of the interactive nature of gathering this data, it's the sort of thing that you can do only during the design phase of running the task, not during run time. The run-time phase provides the SAS program that produces the data and report. If the user wants to update the data to reflect a recent addition or loss of a Facebook friend, then he or she will have to revisit the user interface.

For this task, save the SAS program as the task state. In the custom task API, the task state property is named XmlState. This seems to imply that valid XML is a required format, but XML is simply a convention, not a requirement.

This makes it easy to implement the Get and Restore methods in the SasFacebookTask class (the main custom task class that is located in SasFacebookTask.cs):

```
    string sasProgram; /* will hold saved program */

    public override ShowResult Show(
        System.Windows.Forms.IWin32Window Owner)
    {
```

```
        FacebookConnectorForm dlg =
    new FacebookConnectorForm(true);
    if (dlg.ShowDialog(Owner) ==
        System.Windows.Forms.DialogResult.OK)
    {
        sasProgram = dlg.GetSasProgram();
        return ShowResult.RunNow;
    }
    return ShowResult.Canceled;
}

public override string GetXmlState()
{
    return sasProgram;
}

public override void RestoreStateFromXml(string xmlState)
{
    sasProgram = xmlState;
}

public override string GetSasCode()
{
    return sasProgram;
}
```

Chapter Summary

This chapter is a practical example of using a custom task to connect your SAS process with an external process. As a client application, SAS Enterprise Guide can act as the bridge to bring your external data sources into SAS for analysis.

With a .NET toolkit, you can use the most modern approaches to connect to external services and sources via the web. You can use raw web services if you need to, whether SOAP-based or RESTful in nature. It's even easier when a third party wraps the operation in a .NET code library, as Facebook C# SDK did for me!

Chapter 16: Building a SAS Catalog Explorer

About This Task **219**
   Source Files for This Example 220
About SAS Catalogs **220**
   Working with SAS Catalogs 221
The SAS Catalog Explorer Interaction **225**
   Creating a Utility Window That Floats 226
   Using SAS Workspace APIs 227
   Using Windows Presentation Foundation (WPF) 230
Chapter Summary **231**

Most SAS Enterprise Guide users have no need to peek inside a SAS catalog, which is a file storage medium used internally by many SAS applications. But, long-time SAS programmers might use a SAS catalog as a platform-independent storage medium. They would appreciate a way to look at the catalog and modify the contents. The task example in this chapter shows you how to use data providers to examine SAS file metadata, and how to take advantage of the SAS language to access files in a platform-independent manner.

About This Task

The SAS Catalog Explorer task in this chapter provides a utility to examine the contents of a SAS catalog and perform simple operations, such as delete catalog entries. There are two versions of this task:

- One version is compatible with SAS Enterprise Guide 4.1 and uses Microsoft .NET 1.1 and Windows Forms.

- The other version is compatible with SAS Enterprise Guide 4.2 and 4.3, and uses Microsoft .NET 3.5 and Windows Presentation Foundation (WPF).

Actually, the task version that works with SAS Enterprise Guide 4.1 also works with version 4.2 and 4.3. This compatibility helps illustrate the following important aspects of SAS custom tasks:

Custom tasks are forward compatible. The task that you build to work with one version of SAS Enterprise Guide should continue to work with future versions of SAS Enterprise Guide. As long as your task implementation complies with the documented task API and custom task contracts, it should continue to work with future versions of the application. (If and when a task API becomes deprecated, SAS should provide advance notice and you can plan for it. For example, SAS Enterprise Guide 2.0 custom tasks are supported in SAS Enterprise Guide 3.0 and 4.1, but are deprecated in 4.2).

Building custom tasks using SAS.Tasks.Toolkit results in less code. If you compare the two versions of the SAS Catalog Explorer task (the 4.1 version and the 4.2/4.3 version), you'll notice that the 4.1 version contains a lot more code and is more complex. This is despite the fact that the two versions are almost functionally equivalent.

The 4.2/4.3 version uses WPF for the user interface. WPF is the newer user interface technology from Microsoft. It enables you to encode much of the user interface behavior into a special form of XML called XAML.

With the XAML file format and special WPF features such as data binding, it's easy to present a user interface with a lot of content and a few lines of code. The tricky part is determining exactly what those few lines of code should be.

Source Files for This Example

The task example is named SAS.Tasks.Examples.CatalogExplorer.dll. It is built and ready to use. Here are some details about the task example:

.NET language and version	C# and Visual Basic .NET 1.1 or C# and Visual Basic .NET 3.5 (Microsoft Visual Studio 2003 and 2010)
.NET difficulty	High
.NET features	WPF, data binding, and SAS IOM API
SAS difficulty	High
SAS features	Data access, FILENAME statement
Binaries	SAS.Tasks.Examples.CatalogExplorer.zip
Source code	SAS.Tasks.Examples.CatalogExplorer_src.zip

About SAS Catalogs

SAS catalogs are special files that contain content that is useful in SAS applications. Outside of SAS, a SAS catalog is a single file (usually with a file extension of SAS7BCAT). In SAS, a SAS catalog is like a virtual folder that contains items called entries. There are many different types of SAS catalog entries that are used for different purposes. Some examples of entry types include:

FORMATC
A SAS format definition.

GRSEG
An image created using one of the SAS/GRAPH procedures.

SOURCE
A SAS program. It is often used in a SAS/AF application.

MACRO
A compiled macro program that has already been processed by SAS.

FRAME and SCL
SAS/AF application components. The component is usually part of an interactive application that can be run only in the SAS windowing environment.

Working with SAS Catalogs

Most of the time, SAS automatically creates, modifies, and reads the contents of SAS catalogs without any direct intervention from you. A catalog is a portable file structure that SAS can treat the same on any host (Windows, UNIX, and z/OS), even when those hosts have very different file systems. Because of the portable nature of a SAS catalog, many SAS processes use them behind the scenes.

Using the CATALOG Window

SAS supports several ways for you to see what's inside a SAS catalog. The most intuitive way is using the CATALOG window, which is available only in the SAS windowing environment. Figure 16.1 shows an example of the CATALOG window with a view of the Sashelp.Analyst catalog.

Figure 16.1: The SAS CATALOG Window with Content

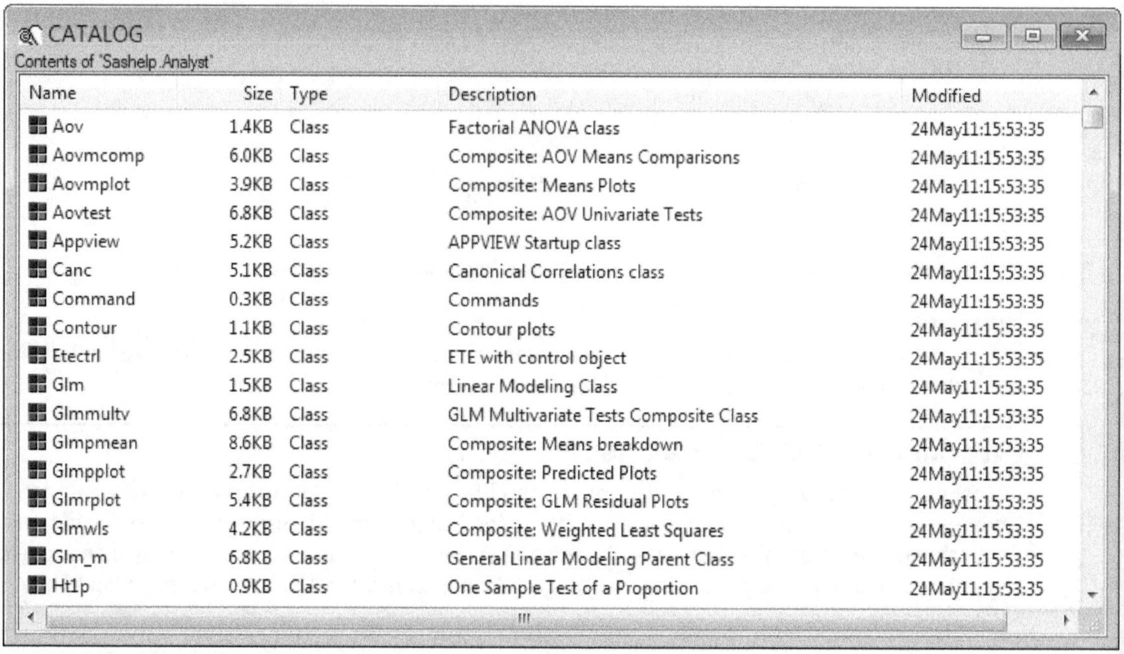
CATALOG
Contents of 'Sashelp.Analyst'
Name
Size
Type
Description
Modified
Aov 1.4KB Class Factorial ANOVA class 24May11:15:53:35
Aovmcomp 6.0KB Class Composite: AOV Means Comparisons 24May11:15:53:35
Aovmplot 3.9KB Class Composite: Means Plots 24May11:15:53:35
Aovtest 6.8KB Class Composite: AOV Univariate Tests 24May11:15:53:35
Appview 5.2KB Class APPVIEW Startup class 24May11:15:53:35
Canc 5.1KB Class Canonical Correlations class 24May11:15:53:35
Command 0.3KB Class Commands 24May11:15:53:35
Contour 1.1KB Class Contour plots 24May11:15:53:35
Etectrl 2.5KB Class ETE with control object 24May11:15:53:35
Glm 1.5KB Class Linear Modeling Class 24May11:15:53:35
Glmmultv 6.8KB Class GLM Multivariate Tests Composite Class 24May11:15:53:35
Glmpmean 8.6KB Class Composite: Means breakdown 24May11:15:53:35
Glmpplot 2.7KB Class Composite: Predicted Plots 24May11:15:53:35
Glmrplot 5.4KB Class Composite: GLM Residual Plots 24May11:15:53:35
Glmwls 4.2KB Class Composite: Weighted Least Squares 24May11:15:53:35
Glm_m 6.8KB Class General Linear Modeling Parent Class 24May11:15:53:35
Ht1p 0.9KB Class One Sample Test of a Proportion 24May11:15:53:35

You can't see this window in SAS Enterprise Guide because built-in SAS windows are not available in the client/server architecture. However, the SAS Catalog Explorer task provides a replacement for the CATALOG window for users who need to see it.

Figure 16.2 shows the SAS Catalog Explorer window with a view of the same Sashelp.Analyst catalog

Figure 16.2: The SAS Catalog Explorer Window with Content

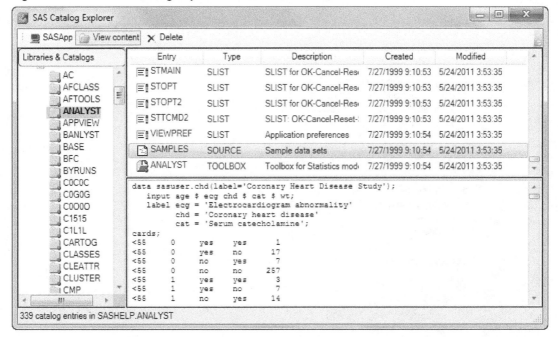

Because you cannot rely on built-in SAS windows for catalog support, you must rely on programmatic methods to access catalog content from within SAS Enterprise Guide. There are three main methods. All of them are used in this custom task. The three methods are the CATALOG procedure, the FILENAME statement with the CATALOG access method, and a SAS DICTIONARY table (SASHELP.VCATALG).

Using the CATALOG Procedure

The CATALOG procedure is a utility procedure that enables you to display catalog contents, rename entries, copy entries among catalogs, and delete entries. In this custom task, the CATALOG procedure deletes an entry when the user clicks the **Delete** button.

Copying entries and renaming them is beyond the scope of this example. If you want to extend the task to support these features, the CATALOG procedure is the way to go.

Here is an example in C# of building a small SAS job to delete a list of SAS catalog entries:

```csharp
private void DeleteCatalogEntries(System.Collections.IList iList)
{
  if (iList.Count == 0) return;

  // log the operation
  _logger.Info("Deleting selected catalog entries");
  // remember which catalog this is so that you can update the
  // list when complete
  SasCatalogEntry e = ((SasCatalogEntry)iList[0]);
  SasCatalog cat = new SasCatalog(Consumer.AssignedServer, e.Libref, e.Member);
  // build PROC CATALOG statements for each catalog entry to
  // delete.
  string code =
    string.Format("proc catalog catalog={0}.{1}; \n",
      e.Libref,
      e.Member);

  foreach (SasCatalogEntry entry in iList)
  {
    code += string.Format("  delete {0}.{1}; \n",
      entry.Entry,
      entry.ObjectType);
  }

  code += "run;\n";

  _logger.InfoFormat("SAS job to delete entries: \n{0}", code);

  // submit the PROC CATALOG job and wait for it to complete
  // this method uses the Submit method from ISASTaskConsumer
  // a bit "old school" compared to using the SasSubmitter
  // class, but it still works
  Consumer.Submit.SubmitCode(code,
    Consumer.AssignedServer, false);

  // update the list view with remaining entries
  UpdateEntriesList(cat);
}
```

Using the FILENAME Statement

The FILENAME statement supports a CATALOG access method, which enables you to treat a catalog entry as a single file in your SAS program. For example, to access the PROGRAM entry named README from the SASHELP.GAMES catalog, use this statement:

```
filename readme catalog "SASHELP.GAMES.README.PROGRAM";
```

The FILENAME statement with the CATALOG access method is used in this task to view the source of text-based catalog entries, such as SOURCE, PROGRAM, and SCL entries. After assigning a file reference using the FILENAME statement, you can use a SAS IOM API (specifically, FileService) to download the content from the SAS workspace to the client.

Later in this chapter, you will learn how to access the SAS IOM APIs in the section **Using SAS Workspace APIs**.

Using a SAS DICTIONARY Table

As examples in other chapters have shown, SAS DICTIONARY tables contain useful metadata about many SAS objects. This includes catalog entries. There is a special VIEW member in the Sashelp library named VCATALG. It contains basic information about a catalog entry, such as the library, catalog name, entry name and type, description, and file timestamps.

To see this for yourself, run the following SAS program from SAS Enterprise Guide:

```
proc sql;
   select objname, objtype, objdesc, created
     from sashelp.vcatalg
     where libname="SASHELP"
           and memname="ANALYST"
  order by objtype;
quit;
```

The output looks very similar to the content shown in Figure 16.2.

If you examine the source for the example that is compatible with SAS Enterprise Guide 4.1, you will see heavy use of the SASHELP.VCATALG content (as well as other members) to gather information for the SAS Catalog Explorer window.

SAS.Tasks.Toolkit, introduced in SAS Enterprise Guide 4.2, contains helpful .NET classes that abstract the metadata about SAS libraries, data sets, catalogs, and catalog entries into objects that are easier to deal with in your .NET code.

In fact, behind the scenes, these helpful .NET classes are using DICTIONARY tables to discover the information. Fortunately, that complexity is hidden from you.

The SAS Catalog Explorer Interaction

The SAS Catalog Explorer task differs from other tasks in this book.

- It doesn't support the normal task cycle of launch, design, and run. Instead, it's a utility window that stays open while you interact with other SAS Enterprise Guide features.
- This task doesn't use traditional Windows Forms technology. Instead, it uses WPF.

- Some parts of this task go beyond what the custom task APIs support. This task requires that you dig deeper into the SAS workspace APIs.

Creating a Utility Window That Floats

Most tasks cannot safely stay visible while you interact with other SAS Enterprise Guide features. The reason is that any task window that enables you to view data or design a report could easily clash with other content that you are accessing in your project.

For example, you cannot safely run a process flow while the SAS Enterprise Guide Query Builder task is displayed. What happens if the task in the process flow tries to update the data being accessed by the Query Builder? The process flow would fail because the Query Builder has a lock on the data in the data set. This is just one of several contention issues that might arise. It's why most SAS Enterprise Guide tasks are designed to be modal.

Modal is the user interface term for a window that stays on top of the current application. A modal window doesn't allow you to interact with the main application until you dismiss the window (by clicking **Cancel** or **OK**). You are in the *mode* of working with that particular window, and you must finish with it before you can move on.

When you allow a window to stay displayed while you interact with other features of the application, the window is considered *modeless*. The SAS Catalog Explorer window is a modeless window.

Making a Task Window Modeless

When you open a traditional task in SAS Enterprise Guide, a few things happen.

- SAS Enterprise Guide creates an instance of your task class and adds it to the active project. You will see the new task icon placed in your process flow, even before you finish using the task.
- SAS Enterprise Guide calls the Show method on your task class.
- SAS Enterprise Guide waits, as if in stasis, until your Show method returns with a result of ShowResult.RunNow, ShowResult.RunLater, or ShowResult.Cancel.
- If the task returns ShowResult.Cancel, then SAS Enterprise Guide removes the task class instance from the active project and leaves no trace that the task was ever invoked.

To trick SAS Enterprise Guide into allowing the task to stay open, do the following:

- In your task class Show method, invoke the task window using a modeless Windows method. In Windows Forms, this is Form.Show instead of Form.ShowDialog. In WPF, use the Window.Show command instead of Window.ShowDialog.
- Using Show instead of ShowDialog returns control immediately to the calling method in your task. And then your task returns ShowResult.Cancel to SAS Enterprise Guide.

Your task window is now visible and SAS Enterprise Guide is no longer waiting for a return code. SAS Enterprise Guide has wiped your task entry from the project. Your task will not be able to save or restore its state in the active SAS Enterprise Guide project (EGP) file.

That's not all there is to it. Because the intent is to create a utility window, you probably don't want the user to be able to invoke the task *again* and create a second (or third or fourth) instance of it. To prevent this, you need to add a static flag that indicates that the task window is currently showing. Then, you need to add logic to prevent it from showing again.

Here's an example from the SAS Catalog Explorer task:

```
// use a static flag to ensure you show only one instance (per process)
internal static bool isOneShowing = false;

// override of the SasTask.Show method
public override ShowResult Show(System.Windows.Forms.IWin32Window
Owner)
{
    if (!isOneShowing)
    {
        isOneShowing = true;
        CatalogExplorerControl c =
            new CatalogExplorerControl(Consumer);
        c.Show();
    }
    return ShowResult.Canceled;
}
```

Don't forget to clear that flag when the SAS Catalog Explorer window is closed by the user. That can be handled in the CatalogExplorerControl class:

```
protected override void OnClosed(EventArgs e)
{
    CatalogExplorer.isOneShowing = false;
    base.OnClosed(e);
}
```

Using SAS Workspace APIs

Before you can use FileService (as required in this task) or any SAS IOM API, you must first add a reference to the SAS Interop assembly in your project. The Interop assembly provides a set of .NET wrapper classes that let you access the features of the SAS IOM API, which is surfaced as a COM object in Windows.

228 *Custom Tasks for SAS Enterprise Guide Using Microsoft .NET*

The SAS Interop assembly is located in the directory in which SAS Enterprise Guide is installed. To add a reference to the SAS Interop assembly in your Visual Studio project:

1. Select **Project→Add Reference**. The Reference Manager dialog box appears.
2. Click the **Browse** button. Navigate to the SAS Enterprise Guide application directory ((for example, `C:\Program Files\SASHome\x86\SASEnterpriseGuide\4.3`).
3. Select **SASInterop.dll**, and click **Add**. Click **OK**. (See Figure 16.3 for an example.) This adds the Interop assembly to your project as an assembly reference.
4. In the Solution Explorer view, expand the list of references, and select SASInterop. Right-click, and select **Properties**. The Properties view shows the current settings for this assembly reference.
5. Change two properties. Set **Copy Local** to **False**, and set **Specific Version** to **False**. By changing these two properties, you'll have less clutter in your task's build directory, and your task will remain compatible with future versions of SAS Enterprise Guide.

Figure 16.3: The Add Reference Window in Visual Studio 2010

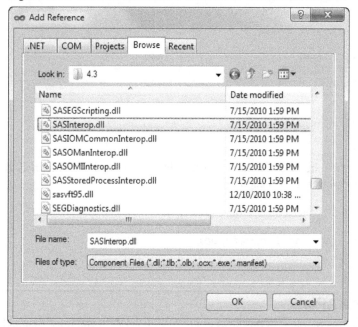

With the reference to the Interop assembly in place, you can now reference the SAS IOM API properties and methods.

Here is an example of using the FileService methods to read a text-based catalog entry into a .NET **string** object:

```
public void ReadEntry(
  ISASTaskConsumer3 consumer,
  string serverName,
  string catEntry)
{

  // you'll be using all of these APIs
  SAS.FileService fs = null;
  SAS.IFileref fr = null;
  SAS.Workspace ws = null;
  SAS.TextStream ts = null;
  string fileref;

  try
  {
    // use SAS workspace and fileservice to download the file
    // get a handle to the SAS workspace interface
    ws = consumer.Workspace(serverName) as SAS.Workspace;
    fs = ws.FileService;

        // using the FILENAME statement CATALOG access method
    fr = fs.AssignFileref("", "CATALOG",
      catEntry, "", out fileref);
    ts = fr.OpenTextStream(
      SAS.StreamOpenMode.StreamOpenModeForReading, 16500);
    ts.Separator = Environment.NewLine;

    StringBuilder sb = new StringBuilder();

    // downloading catalog entry contents
    int lines = 1;
    Array truncLines, readLines;
    // iterate through until all lines are read in
    while (lines > 0)
    {
      ts.ReadLines(100, out truncLines, out readLines);
      lines = readLines.GetLength(0);
      for (int i = 0; i < lines; i++)
        sb.AppendLine(readLines.GetValue(i).ToString());
    }

    txtEntry.Text = sb.ToString();

    // cleanup
    ts.Close();
    fs.DeassignFileref(fr.FilerefName);
  }
```

230 Custom Tasks for SAS Enterprise Guide Using Microsoft .NET

```
    catch (Exception ex)
    {
      MessageBox.Show(
        string.Format(
          "Cannot open the catalog entry {0}.  Reason: {1}",
          catEntry, ex.Message));
    }
  }
```

The flow of the code is this: assign a SAS file reference, read the contents of the file pointed to by the file reference, and then release (clear) the file reference. This flow mirrors the flow that you might follow in a SAS program to achieve the same results, but this API helps you accomplish it without running a SAS program.

Using Windows Presentation Foundation (WPF)

Most of the task examples in this book use Windows Forms in their user interface. This task is different. It uses WPF (just to show that it *can* be done).

In a project that uses WPF, there is a new programming language added to the mix: XAML. XAML is an XML-based description language for user interfaces. You can use an interactive designer in Visual Studio to build much of the XAML code. However, when you need certain effects and behaviors, you will most likely have to tinker with the code by hand.

Figure 16.4 shows the SAS Catalog Explorer window with XAML in the Design view in Visual Studio 2010.

Figure 16.4: SAS Catalog Explorer Window Showing Design View

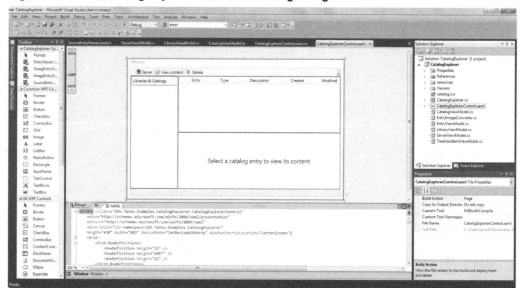

The explorer-style user interface has three main components: a tree view on the left, a grid or list view on the upper right, and a content view on the lower right. In XAML code, you can arrange these main components by containing them in a <Grid> container. When you add the other components (menu items, toolbars, and status bar), you end up with several nested <Grid> objects.

In this book, all of the nuances of using WPF and XAML in your custom tasks aren't covered. The task example helps you learn a few basics. For more information, there are many WPF resources online or in books.

Data Binding

This task example uses data binding. *Data binding* is the practice of populating user interface controls by directing the controls to collections of data objects.

Data binding is extremely useful in .NET programming. It's an essential tactic when developing WPF-based applications. However, data binding can make it challenging to see what's happening in your code.

For example, you might spend hours searching for code that adds libraries and catalogs to the tree view in the SAS Catalog Explorer. There is no such code, really. There is code that populates a set of view-model objects, and then those objects are bound to the tree view with XAML code such as this:

```
<TreeView Grid.Row="1"
  Name="treeView"
  ItemsSource="{Binding Libraries}"
  BorderThickness="0">
```

When something goes wrong with data binding (for example, if the objects are not the expected type), the XAML components are notoriously quiet about it. You won't see obvious exceptions thrown, only incorrect behavior. This can make it tricky to debug data binding issues. It's why XAML programming is almost a sub-discipline unto itself in the world of .NET programming.

Chapter Summary

This chapter touches on many different aspects of custom tasks. First, the SAS Task Toolkit offers built-in objects to access SAS catalogs. These objects are needed only for certain specialized tasks, but they can save you tremendous time and effort when you do need them.

Next, you learned how the raw SAS workspace APIs can be used. The SAS Task Toolkit and custom task APIs already encapsulate much of what you need. When you need to go beyond the built-in features, it's good to know you can.

The example of using WPF illustrates that you are not constrained to the traditional Windows business-style form of a user interface. If you want to offer more advanced levels of interaction for your users, you can use the best of what .NET has to offer.

Chapter 17: Building a SAS Macro Variable Viewer and SAS System Options Viewer

About These Tasks **234**
    Example Source Files and Information 234
Creating a Productive User Experience **234**
    Creating a Toolbox Window 235
    Remembering the Window Position 238
    Adding an About Window with Version Information 241
Designing Code with Objects and Lists **242**
    Planning for Object-Oriented Design 243
    Using Data Structures to Represent SAS Objects 244
Performing Other Cool Tricks **247**
    Checking the Version of SAS 247
    Running SAS Language Functions 248
    Parsing the SAS Log to Detect Line Types 249
Chapter Summary **251**

When you work with a software product long enough, you become very familiar with what it has and what it lacks. You might have an idea for a feature that you think will be really useful. Through its SASware Ballot, SAS Institute regularly receives suggestions for new product features that customers would like to see. Subsequent versions of SAS software often have implementations of these suggestions, although they might take a while to show up. Delayed gratification is better than no gratification at all.

You might have an idea strike you while using SAS Enterprise Guide. Applying the skills that you have learned in this book, you now have the ability to say, "Hey, I don't have to wait for SAS to implement that feature. I can do it myself!"

The two tasks presented in this chapter were born out of the desire for a little extra boost in programmer productivity. This chapter covers two tasks, not just one, because the tasks are similar in their implementations and even share some common code. One task, the SAS Macro Variable Viewer task, enables you to view the current values of your SAS macro variables. The other task, the SAS System Options Viewer task, enables you to view the current values of your SAS system options.

About These Tasks

Because these tasks share some common code, they are implemented in the same Visual Studio project. However, they were not developed at the same time!

The SAS Macro Variable Viewer was developed first. It was made available for download on The SAS Dummy blog (available at http://blogs.sas.com/sasdummy). As users downloaded the task and tried it out, they left comments on the blog with their suggested features. One astute user suggested that it would be useful to be able to view SAS system options as well, similar to what you see in the OPTIONS window in the SAS windowing environment. It was a good suggestion. A month later, an updated version of the task with this capability was available on the blog.

This chapter does not address every detail of these task implementations. Instead, it focuses on just the techniques and tricks that haven't been covered elsewhere in this book. For a complete view of the implementations, download the source projects and review them in Visual Studio.

Example Source Files and Information

Here are some details about the task examples:

.NET language and version	C# and Visual Basic .NET 3.5 (Microsoft Visual Studio 2010)
.NET difficulty	Medium
.NET features	Windows Forms, LINQ, pinvoke, and saving and restoring window preferences
SAS difficulty	Medium
SAS features	DICTIONARY tables, SAS functions, and SAS macro language
Binaries	SAS.MacroViewer.zip
Source code	SAS.MacroViewer_src.zip

Creating a Productive User Experience

Unlike many traditional tasks in SAS Enterprise Guide, these two tasks do not generate a SAS program to run and output results. The user interface *is* the task, so you want to make it as pleasant and productive as possible. Here are a few desired behaviors with brief explanations of each:

Chapter 17: Building a SAS Macro Variable Viewer and SAS System Options Viewer **235**

Behave as a floating toolbox
> Unlike most tasks that are displayed as modal windows (meaning you cannot interact with other features of SAS Enterprise Guide while the task window is showing), you want these tasks to be displayed as modeless windows. A SAS programmer will probably want to edit and run SAS programs and view the macro variables or system options at the same time. This capability is similar to what is provided by the SAS Task Property Viewer from Chapter 10, "For the Workbench: A SAS Task Property Viewer."

Remember its position on the screen, even between uses
> Programmers often like to optimize their use of screen real estate. Because these tasks have always-open windows, a user might want to position the window off to the side or in a second display screen. As a courtesy, the task can record its last window position, and then automatically show itself in that position the next time the user invokes it.

Provide a way to filter its content
> A SAS session might surface hundreds of SAS macro variables and SAS system options. The user needs a simple method to filter the view to specific values of interest.

Enable you to change the selection of the SAS server (for multi-server environments)
> System option settings and macro variable values are different for each SAS session. If you have multiple SAS servers available in your SAS Enterprise Guide session, the task needs to be able to switch your view from one server to another.

Creating a Toolbox Window

If you have used Visual Studio or other similar development environments, then you have worked with toolbox windows. They have certain features and behaviors that you are probably accustomed to, such as:

- A narrower border style than a standard dialog box.
- Ability to float anywhere on the screen (or in a second display screen).
- Ability to resize and have the content layout automatically adjusted in a useful way.
- If appropriate, a toolbar with easy-to-understand icons for common functions.
- If appropriate, a status area to show you a summary of the tool content or current action.

Anatomy of a Toolbox Window

Look at the layout of the SAS Macro Variable Viewer window with an eye toward these features:

Figure 17.1: SAS Macro Variable Viewer as a Toolbox Window

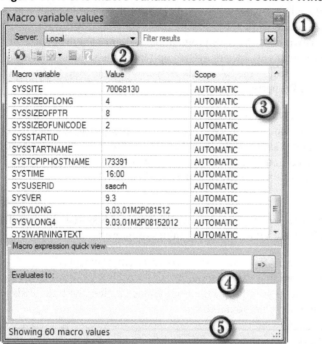

Here are the basic techniques used to implement these features:

- ❶ Like other tasks, this window uses the SAS.Tasks.Toolkit.Controls.TaskForm as a base class. In the Design view in Visual Studio, the FormBorderStyle property is changed to **SizableToolWindow**. This changes the window border to a narrower style.
- ❷ Using the Windows Forms ToolStrip component, you can implement a simple toolbar. The ToolStrip component can host a variety of flexible ToolStripItem components, including a component that acts as a toggle button or a component that acts as a pull-down menu.
- ❸ The window emphasizes the main content. In this window, the main content is the macro variables and their values. A ListView component is used to show the content. It's designed to resize when you change the size of the toolbox, so it should remain as the focal point of the task user interface.
- ❹ For this task, there is an additional feature that enables you to evaluate a macro expression on the fly. It's not the main purpose of the task, so its area takes up less space. You can use one of the buttons in the toolbar to toggle whether this area is displayed. This is achieved by using the Windows Forms SplitContainer component, which makes it easy to expand or collapse certain areas of the window.
- ❺ At the bottom, there is a docked status message area. This is similar to what you've seen in other tasks, such as the SAS Catalog Explorer in Chapter 16, "Building a SAS Catalog Explorer." This area summarizes the content that you see in the main portion of the window.

In this book, I don't describe all of the techniques for designing an effective window layout. When you have a window layout that contains many components and you want a certain behavior for resizing, anchoring, and docking items in the window, it's best to sketch the window on a sheet of graph paper, noting each of the components and how they should be grouped.

Once you have it on paper, annotate the sketch with ideas of which components will use the Anchor or Dock properties and which components should be grouped in a container component such as a Panel or SplitContainer. Windows Forms offers fancier layout components such as TableLayoutPanel and FlowLayoutPanel. These are very flexible components, but you should avoid using more than one or two layers of component nesting or the layout might become difficult to predict and maintain.

Enforcing a Single Instance of the Task

Because the task is taking on a toolbox persona, you don't want the user to launch multiple instances of the window in a single SAS Enterprise Guide session. That could get confusing and the screen could become crowded. What *should* happen when the user selects the SAS Macro Variable Viewer from the menu a second time?

You need to teach the task to run in *Highlander* mode (a nerdy reference to the movie *Highlander*, with its famous quote, "There can be only one."). You want to add a mechanism that checks whether one of these task windows is already showing. If one is, exit the task without launching a second window.

To communicate this status across different instances of the task, use a simple programming technique called a semaphore. A *semaphore* is a flag that shows the state of the toolbox window—is there one showing (true) or not (false)?

In the task class (which inherits from the SAS.Tasks.Toolkit.SasTask class), create a single static Boolean member, and then use it as the semaphore to indicate whether you've got one of these windows already showing. (Remember, a static member has just one value that is shared among all instances of the task.)

In this C# example (located in the MacroTask.cs file), the member is named `isOneShowing`:

```
internal static bool isOneShowing = false;
public override ShowResult Show(System.Windows.Forms.IWin32Window
Owner)
{
    if (!isOneShowing)
    {
        MacroViewerDlg win = new MacroViewerDlg(Consumer,this.Clsid);
        isOneShowing = true;
        win.Show(Owner);
    }
    return ShowResult.Canceled;
}
```

In the task window implementation (which inherits from the SAS.Tasks.Toolkit.Controls.TaskForm class), be sure to clear the semaphore when the window is closed. This snippet is from the MacroViewerDlg.cs file:

```
protected override void OnClosed(EventArgs e)
{
    MacroTask.isOneShowing = false;
    base.OnClosed(e);
}
```

This semaphore affects only a single instance of SAS Enterprise Guide. A user can open multiple instances of SAS Enterprise Guide, and each instance can have its own SAS Macro Variable Viewer window.

Remembering the Window Position

For standard tasks that perform work in your process flow, you use the XmlState property (usually helped by some type of settings class) to store the user settings. Because the tasks in this chapter do not have a representation in your process flow, you need to store the user settings in some other place.

Fortunately, the SAS Task Toolkit provides a helper class for this purpose named SAS.Tasks.Toolkit.Helpers.TaskUserSettings. This class makes it easy to store and retrieve user settings as name and value pairs. The settings are stored in the user's profile area, so the settings are specific to each user. You specify the task ID as a key when you access the settings, so the settings can be specific to just one type of task. All of this means that you don't need to worry about name collisions among different tasks with similar settings.

Storing Your Task's User Settings

In the SAS Macro Variable Viewer and the SAS System Options Viewer, you want to remember the task window's position when the task is closed so that the next time the user opens it, you can place the window in the same position. To achieve this, you need to capture the window's geometry as part of the OnClosing routine, which you can override in the task window class. This C# code is from the BaseToolsForm.cs file in the project:

```
protected override void
OnClosing(System.ComponentModel.CancelEventArgs e)
  {
      SAS.Tasks.Toolkit.Helpers.TaskUserSettings.WriteValue
          (TaskClassID, "XCOORD", Convert.ToString(this.Location.X));
      SAS.Tasks.Toolkit.Helpers.TaskUserSettings.WriteValue
          (TaskClassID, "YCOORD", Convert.ToString(this.Location.Y));
      SAS.Tasks.Toolkit.Helpers.TaskUserSettings.WriteValue
          (TaskClassID, "WIDTH", Convert.ToString(this.Size.Width));
      SAS.Tasks.Toolkit.Helpers.TaskUserSettings.WriteValue
          (TaskClassID, "HEIGHT", Convert.ToString(this.Size.Height));
      SAS.Tasks.Toolkit.Helpers.TaskUserSettings.WriteValue
```

Chapter 17: Building a SAS Macro Variable Viewer and SAS System Options Viewer

```
            (TaskClassID, "DETAILS", Convert.ToString(this.ShowDetails));

        base.OnClosing(e);
    }
```

This code captures the position of the window's upper left corner and its current size. It also captures the toggled state of the detail pane (the macro expression area).

Retrieving Your Task's User Settings

When the task is opened again, the OnLoad routine restores the user settings.

```
    protected override void OnLoad(EventArgs e)
    {
        // check if there are any settings stored for this task
        if (!string.IsNullOrEmpty
              (SAS.Tasks.Toolkit.Helpers.TaskUserSettings.ReadValue
                (TaskClassID, "XCOORD")
              )
           )
        {
            // restore settings, if any, from previous invocations
            try
            {
                int x = Convert.ToInt32
                    (SAS.Tasks.Toolkit.Helpers.TaskUserSettings.ReadValue
                      (TaskClassID, "XCOORD")
                    );
                int y = Convert.ToInt32
                    (SAS.Tasks.Toolkit.Helpers.TaskUserSettings.ReadValue
                      (TaskClassID, "YCOORD")
                    );
                Point p = new Point(x, y);
                if (isPointOnScreen(p))
                {
                    int w = Convert.ToInt32
                        (SAS.Tasks.Toolkit.Helpers.TaskUserSettings.ReadValue
                          (TaskClassID, "WIDTH")
                        );
                    int h = Convert.ToInt32
                        (SAS.Tasks.Toolkit.Helpers.TaskUserSettings.ReadValue
                          (TaskClassID, "HEIGHT")
                        );
                    this.Width = w;
                    this.Height = h;
                    this.Location = p;
                }
```

```
                this.ShowDetails = Convert.ToBoolean
                    (SAS.Tasks.Toolkit.Helpers.TaskUserSettings.ReadValue
                        (TaskClassID, "DETAILS")
                    );
            }
            catch
            { }
        }

        base.OnLoad(e);
    }
```

The OnLoad routine contains several statements that are designed to prevent exceptions. Because restoring the window's position isn't a core function of the task, you want to make sure that if the settings are not stored in the format that you expect them, you don't throw an exception that prevents the task from initializing. The window's position feature is nice to have. If you encounter problems, you can always just allow the task to show itself in the default location.

Ensuring That the Window Is Visible

In that last code example, did you notice the reference to the `isPointOnScreen` method? This little helper method ensures that if the window was last positioned in an area that is now off the screen and not visible, you don't place the window where it will be out of view. This can happen when you have multiple displays during one use of the task and you place the task window on a second display screen. Then, the next time you run the task, you have just a single display (such as when using a remote desktop client).

The `isPointOnScreen` method is a simple check:

```
    bool isPointOnScreen(Point p)
    {
        Screen[] screens = Screen.AllScreens;
        foreach (Screen screen in screens)
        {
            if (screen.WorkingArea.Contains(p))
            {
                return true;
            }
        }
        return false;
    }
```

If the method returns False, then the logic that attempts to place the window in the last known position drops out. The window appears in its default location (which is usually the center of the screen).

Adding an About Window with Version Information

Almost all software applications have an informational window known as the About window. The About window usually contains information about who created the application, links to more information, and legal information such as a copyright notice. The most useful tidbit that you usually find in the About window is the version information.

Because most applications continue to be developed even after their initial release, it's important for a user to know which version he or she is working with. New versions often contain fixes for known problems and sometimes offer new features. Stamping your task with a version number and making the version number easy to discover is an important aspect of supporting your custom task. Figure 17.2 shows an example of the About window for the SAS Macro Variable Viewer.

Figure 17.2: The About Window for the SAS Macro Variable Viewer

In the version information, the first two numbers represent the base version of the task (in this case, **4.3**). The last five numbers are the two-digit year, followed by the three-digit day-of-year count. This file version is a modified version of a Julian date, which is the value that is returned by the SAS JULDATE function. In Figure 17.2, **4.3.0.12267** represents the 267th day of the year 2012 (or September 23, 2012).

It doesn't matter what numbering scheme you choose as long as you increment the version information with each release. You want the user to be able to tell a newer version from an older version. In many SAS applications (such as SAS Enterprise Guide), you'll see this two-digit year and three-digit day-of-year count scheme used.

The version information is embedded into the custom task's DLL file by setting the AssemblyFileVersion property in the AssemblyInfo.cs file:

```
[assembly: AssemblyVersion("4.3.0.0")]
[assembly: AssemblyFileVersion("4.3.0.12267")]
```

The AssemblyVersion property represents the major version of the task. You would typically change this value only when you create a new version that is not compatible with existing versions. The AssemblyFileVersion value becomes part of the DLL file header, so it's the same value that is reported by Windows Explorer when you view the properties of the DLL file.

The following C# code from the AboutDlg.cs file retrieves the version information when the About window is shown:

```
using System.Diagnostics;
public AboutDlg(string taskName)
{
    InitializeComponent();

    // Initialize the name and version label
    lblTaskname.Text = taskName;
    string file =
        System.Reflection.Assembly.GetExecutingAssembly().Location;
    lblVersionNum.Text =
        FileVersionInfo.GetVersionInfo(file).FileVersion;
}
```

Designing Code with Objects and Lists

At the beginning of this chapter, you learned that the two tasks have much in common, including part of the window layout and behavior. When two or more objects have common properties or behaviors, it might make sense to consolidate the common items in a base class. Then, you code each class to inherit from the base class.

Figure 17.3 shows the two tasks side by side in the same SAS Enterprise Guide session.

Chapter 17: Building a SAS Macro Variable Viewer and SAS System Options Viewer 243

Figure 17.3: Two Tasks Similar in Layout and Behavior

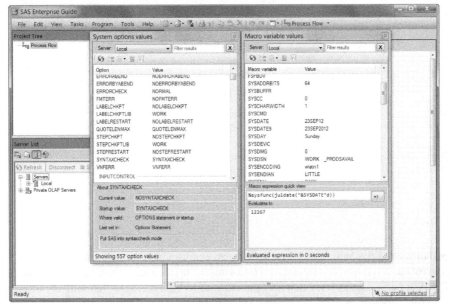

Planning for Object-Oriented Design

The two task windows are different, so it doesn't make sense to implement the behaviors of both windows in the same class. Although that is possible, it makes the tasks difficult to maintain. Whenever you make a change in a class that affects one task's behavior, you have to check the behavior of the other task to make sure you didn't break it.

On the other hand, if you implement the same behavior in two different classes, you do end up with a duplicate set of code. Whenever you make a change in one class, you will probably need to change the other class. You've just multiplied your maintenance effort.

By implementing the common behavior in a base class, you reduce the amount of maintenance in each task class that shares the common behavior. Figure 17.4 shows a class diagram of the BaseToolsForm class, which contains properties and methods that are used by both the SAS Macro Variable Viewer class and the SAS System Options Viewer class. All three of them are derived (directly or indirectly) from SAS.Tasks.Toolkit.Controls.TaskForm, which is provided in the SAS Task Toolkit to implement a few task-specific behaviors.

Figure 17.4: The BaseToolsForm Class and the Two Task Classes

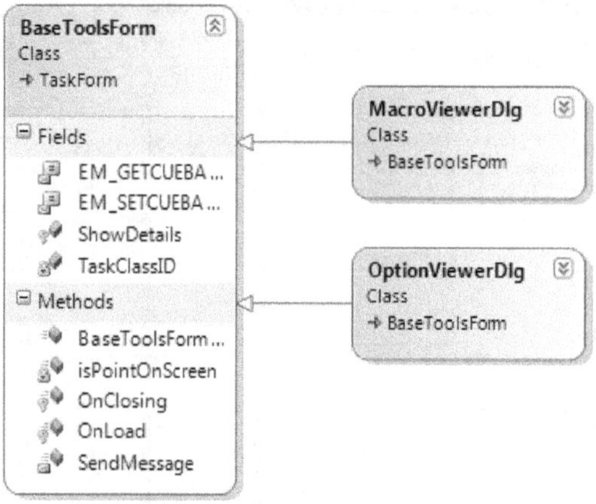

```
BaseToolsForm
Class
 → TaskForm

Fields
    EM_GETCUEBA...
    EM_SETCUEBA...
    ShowDetails
    TaskClassID
Methods
    BaseToolsForm...
    isPointOnScreen
    OnClosing
    OnLoad
    SendMessage

MacroViewerDlg
Class
 → BaseToolsForm

OptionViewerDlg
Class
 → BaseToolsForm
```

The BaseToolsForm class (located in the BaseToolsForm.cs file) implements a few of the shared behaviors in the two tasks. It handles saving and restoring the user settings (including the window size and position). It encapsulates logic that sets the cue text (a little message that tells the user where to type to filter results) in the filter results field.

> *Refactoring* is a software development term that means rearranging your code to accommodate a changing reality. In the case of these two tasks, an entire new task was added after the initial release of the first task, which provided an opportunity to consolidate some of the code into a base class. Constant refactoring is part of the care and feeding of custom tasks or of any code-based project. The Professional Editions of Visual Studio provide several built-in refactoring features such as easy renaming of classes and variables, breaking up large methods into smaller reusable routines, and other operations.

Using Data Structures to Represent SAS Objects

These two tasks show the current values of SAS items—macro variables and SAS system options. They allow the user to interact with these values. You can sort them (by clicking on the column headings), group them (by macro scope or option category), and filter them to show those that match a given pattern.

It's easier to manipulate these values if you create a simple class representation. For example, here's the class structure for tracking a macro variable (located in the MacroVar.cs file):

```csharp
/// <summary>
/// Simple structure to keep the macro variable names/values
/// </summary>
public class MacroVar
{
    /// <summary>
    /// Variable name
    /// </summary>
    public string Name {get; set;}
    /// <summary>
    /// Scope: GLOBAL or AUTOMATIC
    /// </summary>
    public string Scope { get; set; }
    /// <summary>
    /// Current assigned value
    /// </summary>
    public string CurrentValue { get; set; }
}
```

Populating a List of SAS Object Values

With a class structure in place, you can track the current set of values in a .NET generic collection class such as List<>. Here is an example declaration in C#:

```csharp
List<MacroVar> MacroVars { get; set; }
```

You can populate this list by querying the SAS DICTIONARY tables. (For more information, see Chapter 12, "Abracadabra: Turn Your Data into a SAS Program," and Chapter 16, "Building a SAS Catalog Explorer.") In the case of the SAS Macro Variable Viewer, the query applies to SASHELP.VMACRO. This routine is found in the MacroVarsDlg.cs file:

```csharp
/// <summary>
/// Query to retrieve the macro values from the active server
/// </summary>
private void RetrieveMacroValues()
{
    SasServer s = cmbServers.SelectedItem as SasServer;
    using (System.Data.OleDb.OleDbConnection dbConnection =
      s.GetOleDbConnection())
    {
        //----- make provider connection
        dbConnection.Open();

        //----- Read values from query command
        string sql = @"select * from sashelp.vmacro order by name";
        OleDbCommand cmdDB = new OleDbCommand(sql, dbConnection);
```

```
            OleDbDataReader dataReader =
                cmdDB.ExecuteReader(CommandBehavior.CloseConnection);
            try
            {
                // create MacroVar object for each record and
                // add to the list
                MacroVars = new List<MacroVar>();
                while (dataReader.Read())
                {
                    MacroVar var = new MacroVar()
                    {
                        Name = dataReader["name"].ToString(),
                        CurrentValue = dataReader["value"].ToString(),
                        Scope = dataReader["scope"].ToString()
                    };
                    MacroVars.Add(var);
                }
            }
            finally
            {
                dataReader.Close();
                dbConnection.Close();
            }
        }
    }
```

Applying a Filter Using LINQ

With the most recent versions of .NET, Microsoft added a powerful language extension that enables you to filter collections in place using an SQL-like query syntax. The extension is called LINQ. In these tasks, LINQ filters the collection of macro variables or SAS system options in real time as the user types characters in the filter results field. This snippet of code is from the `GetFilteredVars` routine in the MacroVarsDlg.cs file. The routine produces a `List<>` collection of MacroVar objects.

```
    // filter is specified, so match
    // the filter value on the macro name
    // and the macro value
    // using "Contains" for a simple match,
    // but RegEx could provide more flexibility
    var results = from m in MacroVars
        where
            (m.Name.ToUpper().Contains(txtFilter.Text.ToUpper())) ||
            (m.CurrentValue.ToUpper().Contains(txtFilter.Text.ToUpper()))
        select m;
    // populate the List<> with just the filtered results
    foreach (MacroVar mv in results) list.Add(mv);
    return list;
```

Performing Other Cool Tricks

These two tasks contain several useful techniques that you can use in other tasks, including:

- Check the version of SAS so that you can conditionally enable features.
- Execute SAS language functions and retrieve the return values.
- Parse SAS log output to highlight errors and warnings.

This chapter doesn't address all of the techniques used in these tasks. You can study the project to learn more about the additional techniques that interest you.

Checking the Version of SAS

The SAS System Options Viewer task provides information about several aspects of SAS system options, including the value of each option at system start-up time. However, you can retrieve this information only if you have SAS 9.3 or later. Therefore, to keep the task compatible with SAS 9.2 installations, check the SAS version value, and run the task only if it is connected to a SAS 9.3 or later environment.

There are several ways to verify the SAS version. One simple way is to check the value of the SYSVER macro variable. In a custom task, you can retrieve the value of any SAS macro variable by using the `SasServer.GetSasMacroValue` function, available in the SAS Task Toolkit library.

Here is an example (located in the OptionsViewerDlg.cs file) of retrieving and parsing the SAS version value:

```
SasServer s = cmbServers.SelectedItem as SasServer;
double dServerVer = 9.2;
string ver = s.GetSasMacroValue("SYSVER");
try
{
    dServerVer = Convert.ToDouble(ver);
}
catch { }
```

The return value of `GetSasMacroValue` is always a text string. To use the value in numeric comparisons, you need to convert the value to a decimal value (a .NET `Double` type in this case). Although the value should always be a valid decimal number, the `Convert.ToDouble` routine can throw an exception if it's not valid, so a simple Try and Catch block is added around the operation.

The Catch statement catches all exceptions, which isn't really the best practice. But, in this case, the fallback value is initialized to 9.2. If a problem occurs, the task won't halt. Instead, it degrades to 9.2 compatibility.

Running SAS Language Functions

The SAS APIs do not provide a direct method for executing a SAS language function and retrieving the results in your custom task. Instead, there is a technique that involves multiple steps.

1. Build a short SAS program that executes the SAS language functions that you want to run using the `%SYSFUNC` macro function.
2. In the SAS program, assign the value of a function result to a SAS macro variable.
3. Use the `SasSubmitter` class to run the SAS program in the appropriate SAS server environment.
4. Use the `SasServer.GetSasMacroValue` function to retrieve each result value.

In the SAS System Options Viewer task, use the previous steps to access the GETOPTION function in SAS. GETOPTION retrieves details about a specific SAS system option. Here is an example from the OptionsViewerDlg.cs file:

```
string optionDetails =
    "%let _egopthowset = %sysfunc(getoption({0},,howset)); \n" +
    "%let _egoptfullval = %sysfunc(getoption({0})); \n";
string optionStartup =
    "%let _egoptstartup = %sysfunc(getoption({0},,startupvalue));";
/// <summary>
/// Get the details about the SAS system option
/// and its value
/// </summary>
/// <param name="optionName">Name of the option</param>
/// <param name="whereset">return value: where the value was set</param>
/// <param name="startup">return value: value at startup</param>
/// <param name="fullval">return value: full value of the option</param>
private void getOptionInfo(string optionName,
    ref string whereset,
    ref string startup,
    ref string fullval)
{
    whereset = "<cannot determine>";
    startup = "<cannot determine>";
    try
    {
        SAS.Tasks.Toolkit.SasSubmitter submitter =
            new SasSubmitter(cmbServers.Text);
        if (!submitter.IsServerBusy())
        {
            string outLog;
            string program = string.Format(
                optionDetails +
                (dServerVer > 9.2 ? optionStartup : ""),
                optionName);
```

```
                    bool success =
                      submitter.SubmitSasProgramAndWait(program, out outLog);
                    if (success)
                    {
                        SasServer s = new SasServer(cmbServers.Text);
                        whereset = s.GetSasMacroValue("_EGOPTHOWSET");
                        fullval = s.GetSasMacroValue("_EGOPTFULLVAL");
                        if (dServerVer > 9.2) startup =
                          s.GetSasMacroValue("_EGOPTSTARTUP");
                    }
                }
            }
            catch
            {
            }
        }
```

Did you notice that this example contains logic to check whether the SAS server version is higher than 9.2? If the SAS session is running SAS 9.2, the code omits the check for **startupvalue** because that value is not supported by the GETOPTION function in versions earlier than SAS 9.3.

These calls to the SAS APIs are wrapped in a Try and Catch block as a defensive measure with the Catch statement poised to catch all exceptions. In programmer-speak, this is jokingly called the *Pokémon* approach to exception handling (when you just gotta catch them all).

Parsing the SAS Log to Detect Line Types

The SAS Macro Variable Viewer has a nifty feature that enables you to evaluate a SAS macro value or expression instantly and directly in the task window. This technique is similar to the technique that you used to retrieve SAS system option details in the SAS System Options Viewer. You submit a short SAS program that assigns the result to a SAS macro variable, and you then use SasServer.GetSasMacroValue to retrieve the value of that macro variable.

Because this feature provides a text field in which the user can enter any expression, there is a greater possibility for an error to occur. You want to report errors and warnings in the results text box. As an added feature, you can color-code ERROR and WARNING lines in the same colors that the user sees them in in a SAS program log on SAS Enterprise Guide.

To achieve these results, you need to parse the SAS log line by line to detect whether a line represents a warning or error message. It's not enough to simply look for the words "ERROR" or "WARNING" for these reasons:

- When you run SAS in a non-English language, the words "ERROR" and "WARNING" might be translated into the current language.
- A single error or warning message might occupy multiple lines in the log output. If you attempt to detect the line type by looking for a certain word, you might catch the first line of a message but not the subsequent lines.

Fortunately, each line of the SAS log output is tagged with a two-byte line-type prefix, which provides a reliable cue for detecting it programmatically. Error lines start with the e character. Warning lines start with the w character. In the SAS log viewer, these prefix characters are stripped out before they are displayed so that you see only the substance of the SAS log.

Here is the code (located in the MacroViewerDlg.csfile) that evaluates a SAS macro expression and populates the results text box with either the results or errors and warnings:

```
// submit expression and wait
bool success = submitter.SubmitSasProgramAndWait(resolveProgram, out outLog);
if (success)
{
    SasServer s = new SasServer(sasServer);
    txtEval.Text = s.GetSasMacroValue("_EGMACROQUICKEVAL");

    // scan output log for warning messages
    string[] lines = outLog.Split('\n');
    string msg = "";
    foreach (string l in lines)
    {
        // WARNING lines prefixed with 'w'
        if (l.StartsWith("w"))
        {
            // color WARNINGs blue
            msg = msg + l.Substring(2) + Environment.NewLine;
            txtEval.ForeColor = Color.Blue;
        }
    }
    txtEval.Text += Environment.NewLine + msg;
}
else
{
    // scan output log for error lines
    string[] lines = outLog.Split('\n');
    string msg = "";
    foreach (string l in lines)
    {
        // ERROR lines prefixed with 'e', WARNING with 'w'
        if (l.StartsWith("w") || l.StartsWith("e"))
            msg = msg + l.Substring(2) + Environment.NewLine;
    }
    // color ERRORs red
    txtEval.Text = msg;
    txtEval.ForeColor = Color.Red;
}
```

Chapter Summary

As the final chapter in this book, the techniques described in this chapter represent a culmination of many of the topics that were addressed in earlier chapters. The examples feature cool uses of SAS services, such as data access and program execution. They show off some advanced Microsoft .NET programming techniques for creating responsive user interfaces that are easy to manage.

The two examples presented in this chapter are designed to increase productivity by providing convenient features to SAS programmers and SAS Enterprise Guide power users. You can probably think of other features that might fit a similar task model, surfacing just the information that your SAS users need when they need it.

Index

A

About window 241–242
access control to custom tasks 10–11
Add-In Manager dialog box 9
add-ins
   See custom tasks
Add New Item dialog box
   using Visual Basic 75, 78
   using Visual C# 90, 93
Add References dialog box 84, 228
ADO.NET 166–168
APIs
   See interfaces
%appdata% (environment variable) 7–8, 107
asynchronous programming methods 216

B

BeginInvoke method 216–217
binding data 188, 231
breakpoints 111, 115

C

C# language
   about 15
   adding code to tasks 85–88
   adding user interface to tasks 89–92
   building, deploying, testing custom tasks 89
   creating custom task projects using 32–35
   creating custom tasks using 83–96
   creating projects 84–85
   saving and restoring task settings 92–96
   usage considerations 17–18
calculating running totals
   about 141–142
   across groups 144–145
   assumptions for 143
   designing task features 142–145
   designing user interface 146–156
   for one measure across all rows 143–144
   generating SAS programs 160–163
   saving user selections 157–160
CATALOG procedure 223–224
CATALOG window 221–223
character variables example
   about 177–178
   adapting samples 178–182
   adding reporting 180–182
   modeling options 187–189
   running and repeating tasks 189–190
   wrapping samples in tasks 183–190
Choose Toolbox Items dialog box 148–149, 170–171
CIL (Common Intermediate Language) 15
class inheritance 98–99
CLASS statement, MEANS procedure 121
CLR (Common Language Runtime) 15
CodePlex website 24
CodeProject website 23–24
columns, reading and showing 124–127
combo boxes 188–189
commands, running system
   about 193–195
   implementing 196–201
   structure of 195
Common Intermediate Language (CIL) 15
Common Language Runtime (CLR) 15
COMPRESS= data set option 179, 187–188
compression methods
   about 179–180
   reporting results of 180–182
CONTENTS procedure 181
CONTENTS statement, DATASETS procedure 181
controls
   adding 148–151
   hooking to data and events 151–156
   reverse engineering example 170–172
   variable selector 147, 152–155
CSPROJ files 111

254 Index

custom task APIs
   See interfaces
custom tasks
   about 1–2
   access control to 10–11
   accessing from Add-In menu 9–10
   adding Windows Forms to 75–77
   building 74, 89
   building SAS Catalog Explorer 219–232
   building SAS Macro Variable Viewer 233–251
   building SAS System Options Viewer 233–251
   building SAS Task Property Viewer 131–139
   calculating running totals 141–163
   connecting to Facebook example 203–218
   creating 3–4
   creating elegant flow 169
   creating Top N report 117–129
   creating using Visual Basic 30–32, 69–81
   creating using Visual C# 32–35, 83–96
   creating using Visual Studio 27–28, 35–39
   debugging 112–116
   deploying 6–12, 74, 89
   downloading collection of 12
   evolution of 5–6
   example uses for 4–6
   limitations of 5
   retrieving in process flows 135–137
   reverse engineering example 165–176
   running and repeating 189–190
   saving and restoring settings 77–81, 92–96
   saving results 217–218
   special interfaces for 49–53
   stamping with version number 241–242
   system command example 193–201
   task-specific filters 161–162
   testing 74, 89
   Visual Studio versions and 19

D

data binding 188, 231

data models 213
data sets
   dissecting 167
   reading data from 167–168
   reverse engineering example 165–176
   shrinking character variable lengths 177–191
data sources, accessing 59–60
DATA step
   calculating running totals 143–145
   Facebook example 205–206, 213, 215–216
   reverse engineering example 165–176
   Top N report example 121
DATASETS procedure 181
DATE5. format 207
debugging techniques
   about 97
   best practices for 98–111
   for projects 21
   handling exceptions 100–105
   logging mechanism and 106–111
   object-oriented design and 98–99
   troubleshooting breakpoints 115–116
   unit testing 99–100
   using Visual Studio 111–116
delegate keyword 217
deploying custom tasks
   about 6
   Add-In Manager dialog box 9
   common questions about 10–11
   drop-in deployment method 7–9
   using Visual Basic 74
   using Visual C# 89
DICTIONARY tables 166–167, 181, 225, 245–246
disassemblers 22
discovery phase 45
DLL files (.NET assemblies)
   about 111
   accessing 10–12
   calculating running totals example 147–148
   tools for 21–22

DOLLAR*w*. format   127
drop-in deployment method   7–9

E

EGP files   227
encapsulation   98, 166
ERROR message   198, 249
exception handling
    about   100–101
    "invalid cross-thread operation" exception   216
    preventable exceptions   101–102
    recoverable exceptions   102–104
    testing   104–105

F

Facebook API   204, 209–210
Facebook example
    about   203–204
    analyzing data   206–210
    connecting to   203–218
    gathering data   205–206
    main steps for building   211–218
    running   210
    saving results   217–218
File dialog box   17–18, 184–185
FILENAME statement   224–225
FileService API   225, 227–230
filters
    applying using LINQ   246
    task-specific   161–162
FilterSettings class   161
FIRST. automatic variable   144
FORMATC entry type   221
FormattedLogWriter class   175, 198
FRAME entry type   221
FromXml() function   159, 189

G

garbage collection   15
GCHART procedure   121
GETOPTION function   248
GitHub website   24

groups, calculating running totals across   144–145
GRSEG entry type   221
Guidgen tool   71, 85

H

handling exceptions
    about   100–101
    "invalid cross-thread operation" exception   216
    preventable exceptions   101–102
    recoverable exceptions   102–104
    testing   104–105
Highlander mode   237

I

IL (Intermediate Language)   15
Implements keyword   196
Import Data task   194
installing project templates   29–30
interfaces
    about   41–42
    for special tasks   49–53
    required for custom tasks   43–44
    versioned   43–44
Intermediate Language (IL)   15
"invalid cross-thread operation" exception   216
ISASApplicationOptions interface   52
ISASProject interface
    about   49, 51–52, 132
    accessing properties with   134–137
    ProcessFlows property   51
ISASProjectItem interface   51, 137
ISASProjectProcessFlow interface
    about   49, 51–52, 132, 138
    AddLink method   138
    AreLinked method   138
    CanLink method   138
    custom task example   132, 135–136
    GetAllSasCode method   138
    GetSasCodeForThisPath method   138
    GetSasCodeUpToHere method   139
    GetTasks() method   51

ISASProjectProcessFlow interface (*continued*)
    IsAncestor method 138
    RemoveLink method 138
ISASProjectTask interface
    about 51
    custom task examples 132, 137
    XmlState property 51, 132
ISASTask interface
    about 43
    Initialize() method 46–47
    SasCode() method 47
    Show() method 46
    Terminate() method 46–47
    XmlState() method 46–47
ISASTask3 interface
    about 44
    OutputDataDescriptorList method 47
ISASTaskAddIn interface
    about 43
    Connect() method 45–47
    Disconnect() method 46–47
ISASTaskConsumer interface
    about 43, 45
    ShowOutputDataSelector method 185
    Top N report example 127
ISASTaskConsumer3 interface
    about 44
    GetApplicationOptions() method 52
    GetProject() method 52, 135
    GetThisProcessFlow() method 52
    GetThisTask() method 52, 136
ISASTaskData interface 43
ISASTaskDataAccessor interface
    about 43
    Top N report example 127
ISASTaskDataColumn interface 43
ISASTaskDataDescriptor interface 44
ISASTaskDescription interface
    about 43
    FriendlyName property 45
    Languages() method 45
    Languages property 45
    TaskName property 45
    WhatIsDescription property 45
ISASTaskExecution interface
    about 49–51
    Cancel() method 173, 196–197
    OpenResultStream method 196–197
    ResultCount property 174, 196–197
    reverse engineering example 166, 172–175
    Run method 174–175, 196–197, 201
    system command example 196–201
    usage considerations 201
ISASTaskLogWriter interface 175
ISASTaskSubmit interface 43
ISASTaskTemplate interface 52–53
ISupportedRoleBased interface 53
ITaskStream interface 174

J

JIT (just-in-time) compilation 15
JSON (JavaScript Object Notation) 204–206

L

language-integrated query (LINQ) 157–160, 246
LENGTH statement 178
librefs 185
life cycle of tasks
    about 42, 44–45
    during discovery phase 45
    during open and show phase 45–46
    during run phase 46–47
    observing 47–49
LINQ (language-integrated query) 157–160, 246
ListView component (Windows Forms) 236
Log Task API Calls dialog box 48
logging mechanism (log4net)
    about 11, 106–107
    adding loggers 107–109
    detecting line types 249–250
    enabling 107
    profiling tasks 109–111
    system command example 198
logging.config file 107

M

log4net.dll file 107

MACRO entry type 221
Macro Variable Viewer task 98–99
macro variables
    retrieving values of 66
    SAS Macro Variable Viewer task 233–251
macros, refactoring 178–179
managed programming models 15
marshalling code 216
MEANS procedure
    CLASS statement 121
    Top N report example 121
MethodInvoker delegate 217
Microsoft Intermediate Language (MSIL) 15
Microsoft .NET Framework
    See .NET Framework
modal windows 226, 235
modeless windows 226–227, 235
MSIL (Microsoft Intermediate Language) 15

N

.NET assemblies (DLL files)
    about 111
    accessing 10–12
    calculating running totals example 147–148
    tools for 21–22
.NET Framework
    about 15–16
    best practices for debugging software 98–111
    object-oriented design and 98–99, 243–244
    programmer considerations 23–24
    programming language considerations 17–18
    third-party components 24
    tools supported 21–22
    version considerations 84
New Project dialog box
    using Visual Basic 30, 32, 70
    using Visual C# 33, 35

using Visual Studio 35–38
NOTE message 198
NotImplementedException 196–197
null reference exceptions 101–102, 104–105

O

obfuscators 22
object-oriented design 98–99, 243–244
objects, data structures representing 244–246
OLE DB protocol 167
open and show phase 45–46
OpenFileDialog class 18
Options Value Viewer task 98–99
OverflowException 105

P

PDB file (symbols file) 111
platform independence 15
polymorphism 98
PowerShell 195
Process class 200
processes
    attaching debuggers to 112–113
    retrieving flows 135
    retrieving tasks in 135–136
    stored 3–4
    system command example 193–201
ProcessStartInfo class 201
profiling
    about 109–111
    tools supporting 21–22
programs
    See SAS programs
project templates
    about 27–28
    creating custom tasks using 28–39
    downloading 28
    installing on PCs 29–30
projects
    See also custom tasks
    about 20–21
    creating using Visual Basic 30–32, 70–81
    creating using Visual C# 32–35, 84–96

258 *Index*

projects (*continued*)
    creating using Visual Studio 35–39
    source 111
prompts, adding to SAS programs 2–4
properties
    accessing using ISASProject interface 134–137
    displaying in user interface 133–134
    retrieving for selected tasks 137
Properties dialog box
    Debug tab 114
    for projects 32, 35, 37

Q

Query Builder 226
quotation marks (") 163

R

Red Gate Software Reflector tool 22
refactoring macros 178–179
Reference Manager dialog box 70–71, 108, 228
reflection feature 22, 129
Reflector tool (Red Gate Software) 22
regaddin.exe utility 45
REPORT procedure 121
reporting
    compression results 180–182
    Facebook example 207–209
REST (representation state transfer) 210
ResultInfo class 174–175
reverse engineering example
    about 165–166
    adding SAS Enhanced Editor to Windows Forms 170–172
    creating elegant task flow 169
    dissecting data sets 167
    reading data from data sets 167–168
rows, calculating running totals across 143–144
run phase 46–47
running totals, calculating
    about 141–142
    across groups 144–145
    assumptions for 143
    designing task features 142–145
    designing user interface 146–156
    for one measure across all rows 143–144
    generating SAS programs 160–163
    saving user selections 157–160

S

SAS Add-In for Microsoft Office
    about 15
    drop-in deployment method 7–8
    interface support 41–42, 44
    process names 112
SAS/AF 5
SAS Catalog Explorer task
    about 219–220
    about SAS catalogs 220–225
    building 219–232
    creating utility window that floats 226–227
    interaction in 225–231
SAS catalogs
    about 219–221
    working with 221–225
SAS data sets
    See data sets
SAS DICTIONARY tables 166–167, 181, 225, 245–246
SAS Enhanced Editor 170–172
SAS Interop assembly 227–228
SAS IOM API 227–228
SAS IOM OLE DB data provider 167
SAS language functions 248–249
SAS Macro Variable Viewer task
    about 233–234
    creating productive user experience 234–242
    designing code with objects and lists 242–246
    useful techniques 247–250
SAS objects, data structures representing 244–246
SAS Program Runner dialog box 115
SAS programs
    adding prompts to 2–4

attaching debuggers to   112–113
calculating running totals example   160–163
creating Top N report   121–123
Facebook example   215–216
interfaces and   41
reverse engineering example   165–176
writing   2, 4
SAS System Options Viewer task
  about   233–234
  creating productive user experience   234–242
  designing code with objects and lists   242–246
  useful techniques   247–250
SAS Task Property Viewer
  about   131–132
  custom task example   132–139
SasCatalog class   60
SasCatalogEntry class   60
SASCodeViewDialog class   63
SasColumn class   59
SasData class
  about   59
  calculating running totals example   154
  GetColumns() method   102–104
SasDataException   103
SAS.EG.Controls.dll file   147, 149
SasLibrary class   59, 64–66
SASLogViewDialog class   63, 66–67
SASMETADATAREQUIRED attribute   11
SasServer class
  about   59
  GetSasMacroValue method   247–249
  reverse engineering example   167
  usage example   64–66
SAS.Shared.AddIns namespace   55, 132
SAS.SharedUI.dll file   147
SasSubmitter class
  about   60
  IsServerBusy method   60
  SubmitSASProgram method   61

SubmitSASProgramAndWait method   60–61
SasTask class
  about   56–59
  extending   196–198
  GeneratesSasCode property   197
  GetSasCode() method   57
  GetXmlState() method   57
  RestoreStateFromXml() method   57
  Show() method   57
  usage example   72–73, 85–86
SAS.Tasks.Toolkit namespace
  about   55–56
  accessing data sources   59–60
  class usage examples   63–67
  interface support   41
  SasServer class   59, 64–66, 167, 247–249
  SasSubmitter class   60–61
  SasTask class   56–59, 72–73, 85–86, 196–198
SAS.Tasks.Toolkit.Controls namespace
  about   62
  SASTextEditorCtl class   63, 166, 170–172
  TaskForm class   99, 236, 238
SAS.Tasks.Toolkit.Data namespace
  about   59
  SasCatalog class   60
  SasCatalogEntry class   60
  SasColumn class   59
  SasData class   59, 102–104, 154
  SasLibrary class   59, 64–66
SAS.Tasks.Toolkit.Helpers namespace
  about   61
  FormattedLogWriter class   175, 198
  ResultInfo class   174–175
  TaskAddInHelpers class   62
  TaskDataHelpers class   61–62
  TaskOptions class   62
  UtilityFunctions class   62, 163, 190
SAS.Tasks.Toolkit.SasTask file   32, 35, 37
SASTextEditorCtl class   63, 166, 170–172
SASware Ballot   233
Save Project dialog box   31, 34

Scatter Plot Matrix task 160–161
SCL entry type 221
SEGuide.exe process 112
semaphore, defined 237
serializable classes 127–129
SIN files 111
social media, connecting to Facebook 203–218
SORT procedure 121
SOURCE entry type 221
source projects 111
SourceForge website 24
SplitContainer component (Windows Forms) 236–237
static code analyzers 22
stored processes 3–4
String class 18
symbols file (PDB file) 111
%SYSFUNC macro function 248
SYSTASK statement 193–195, 201
system command example
    about 193–195
        implementing 196–201
        structure of 195
system options, viewing values of 233–251
System.Collections.Generic namespace 215
System.Diagnostics namespace 200–201
System.Text.RegularExpressions.Regex class 156
SYSVER macro variable 66

T

task-specific filters 161–162
Task Toolkit library
    See SAS.Tasks.Toolkit namespace
TaskAddInHelpers class
    about 62
    CreateTaskTemplate() method 62
TaskDataHelpers class
    about 61–62
    GetDistinctValues() class 61
    GetSasCodeReference() method 61
TaskForm class 99, 236, 238
TaskOptions class
    about 62
    GetDefaultFootnoteText() method 62
TDD (test-driven design) 100
test-driven design (TDD) 100
testing custom tasks
    for exceptions 104–105
    unit testing 99–100
    using Visual Basic 74
    using Visual C# 89
toolbox windows
    anatomy of 235–237
    creating 235–238
ToolStrip component (Windows Forms) 236
ToolStripItem component (Windows Forms) 236
Top N report
    about 117–118
    creating custom task 123
    creating SAS program 121–123
    examining solution 123–129
    exploring problem 119–121
    reading and showing available columns 124–127
ToXml() function 157, 189
troubleshooting breakpoints 115–116
Try and Catch block 103–104, 247, 249

U

underscore (_) 71
unit testing 99–100
user interfaces
    accessing properties with ISASProject interface 134–137
    adding with Visual Basic 75–77
    adding with Visual C# 89–92
    calculating running totals example 146–156
    character variables example 183–184
    displaying properties in 133–134
    Facebook example 212–213, 216–217
    system command example 195
UtilityFunctions class
    about 62

GetValidSasName() method   62
ReadFileFromAssembly() method   190
SASValidLiteral() method   62, 163
uuidgen.exe utility   71, 85

V

validating user input   156, 185–187
VALIDVARNAME= system option   162
variable selector control   147, 152–155
variables
   shrinking lengths of   177–191
   wrapping names   162–163
VBPROJ files   111
version element (XML)   158
versions
   checking for SAS   247
   considerations for .NET Framework   84
   for interfaces   43–44
   selecting for Visual Studio   18–20
   stamping custom tasks with   241–242
Visual Basic .NET
   about   15
   adding code to tasks   71–74
   adding user interface to tasks   75–77
   building, deploying, testing custom tasks   74
   creating custom task projects   30–32
   creating custom tasks using   69–81
   creating projects   70–71
   Implements keyword   196
   saving and restoring task settings   77–81
   usage considerations   17–18
Visual C#
   adding code to tasks   85–88
   adding user interface to tasks   89–92
   building, deploying, testing custom tasks   89
   creating custom task projects   32–35
   creating custom tasks using   83–96
   creating projects   84–85
   saving and restoring task settings   92–96
Visual Studio
   about   18

creating custom tasks   27–28
creating custom tasks using project templates   28–39
debugging with   111–116
development environment for   20–21
selecting a version   18–20

W

WARNING message   198, 249
websites for additional resources   23–24
WHERE= data set option   161
window positions   235, 238–240
Windows Forms
   adding SAS Enhanced Editor to   170–172
   adding to tasks using Visual Basic   75–77
   adding to tasks using Visual C#   89–92
   ListView component   236
   SAS Catalog Explorer task   227
   SplitContainer component   236–237
   ToolStrip component   236
   ToolStripItem component   236
   WIDTH property   105
WPF (Windows Presentation Foundation)   23, 225, 230–231

X

x command   193–194, 201
XAML file format   220, 230–231
XDocument class (LINQ)   157–160
XElement class (LINQ)   157–160
XmlSerializer class   128–129
XSYNC system option   201

Numbers and Symbols

_ (underscore)   71
" (quotation marks)   163

ACCELERATE YOUR SAS® KNOWLEDGE WITH SAS BOOKS

Learn about our authors and their books, download free chapters, access example code and data, and more at **support.sas.com/authors**.

Browse our full catalog to find additional books that are just right for you at **support.sas.com/bookstore**.

Subscribe to our monthly e-newsletter to get the latest on new books, documentation, and tips—delivered to you—at **support.sas.com/sbr**.

Browse and search free SAS documentation sorted by release and by product at **support.sas.com/documentation**.

Email us: sasbook@sas.com
Call: 800-727-3228

THE POWER TO KNOW®

SAS Institute Inc. product or service names are registered trademarks or trademarks of SAS Institute Inc. in the USA and other countries. ® indicates USA registration. Other brand and product names are trademarks of their respective companies. © 2012 SAS Institute Inc. All rights reserved. S93150US.0612